T0178565

Health and Safety Management

An Alternative Approach to Reducing Accidents, Injury and Illness at Work

Health and Safety Management

An Alternative Approach to Reducing Accidents, Injury and Illness at Work

John White

CRC Press
Taylor & Francis Group
Boca Raton London New York

CRC Press is an imprint of the
Taylor & Francis Group, an **informa** business

CRC Press
Taylor & Francis Group
6000 Broken Sound Parkway NW, Suite 300
Boca Raton, FL 33487-2742

© 2019 by Taylor & Francis Group, LLC
CRC Press is an imprint of Taylor & Francis Group, an Informa business

No claim to original U.S. Government works

Printed on acid-free paper

International Standard Book Number-13: 978-1-138-50083-9 (Paperback)
978-1-138-50084-6 (Hardback)

Library of Congress Cataloging-in-Publication Data

Names: White, John, 1955- author.
Title: Health and safety management : an alternative approach to reducing accidents, injury and illness at work / John White.
Description: Boca Raton : Taylor & Francis, CRC Press, 2018.
Identifiers: LCCN 2018014392| ISBN 9781138500839 (pbk.) | ISBN 9781138500846 (hardback : alk. paper) | ISBN 9781315144023 (ebook)
Subjects: LCSH: Industrial safety. | Accidents--Prevention.
Classification: LCC T55 .W497 2018 | DDC 658.3/82--dc23
LC record available at https://lccn.loc.gov/2018014392

Visit the Taylor & Francis Web site at
http://www.taylorandfrancis.com

and the CRC Press Web site at
http://www.crcpress.com

Contents

Preface

Statistics repeatedly tell us that human error contributes directly to approximately 80% of workplace accidents, incidents and near misses, while statistics from the Health and Safety Executive (HSE) illustrate a steady stagnation in the rate of reduction of accidents, injuries and time taken off work due to ill health. At a time when our awareness and application of health and safety strategies and policies, together with health and safety culture, are at their most advanced state, we should perhaps collectively expect the statistics to reflect this through a higher rate of reduction.

In order to continue to achieve a reduction in the accident statistics, I strongly believe that alongside providing effective systems, procedures and training, it is essential to explore ways of managing the human dynamic more effectively. This belief is backed up by my conversations with several leading figures in major construction projects and my experiences in managing high-risk outdoor activities over a 25-year period.

By applying a range of new and existing principles, including free-thinking hazard identification and marginal gain and using lessons drawn from sport and the high-risk world of outdoor adventure, this book provides a variety of practical solutions that will in turn:

- Reduce the incidence of human error in the workplace
- Reduce the number of accidents, incidents and near misses
- Reduce time taken off work due to injury or ill health

The benefits of such an approach are good for the individual, families, healthcare system, employer and society in general.

Author

Following training in landscape management, John took up various positions with the Lake District National Park Authority, including Volunteers Coordinator and National Park Ranger for the busy central Lake District. After qualifying as a lecturer in countryside management, he went on to set up and run a successful mountain guiding business, which led to a variety of work with colleges, universities and a range of businesses. John also developed Highpoint and Lakes Challenge, a business providing team- and relationship-building events for a wide range of clients including English Partnerships, The Olympic Delivery Authority, Tideway and Loughborough University. He also worked as Director of Conservation and Development for a leading conservation organization.

John is an accomplished climber with over 100 first ascents to his name from Africa to North America, and he has led expeditions to Greenland alongside climbing throughout the Alps and in many European countries. A keen naturalist, he was extensively involved in birds of prey protection schemes and retains a long-standing interest in the natural world.

John has managed health and safety in many difficult and extreme environments and sports and has been responsible for the safety of many thousands of participants in outdoor events. He set up a detailed health and safety policy for a leading conservation organization and was responsible for the safe working of hundreds of volunteers alongside full-time staff.

Previous works by the author include:

Rock Climbing in Northern England
Rock Climbing in Langdale
AA Classic Walks and Tours in France
AA Secret France
AA Touring France
Classic Walks in Great Britain
Classic Walks of the World
(All as contributing author)
Walking in the Stubai Alps
The Indoor Climbing Manual

John has also written extensively for outdoor magazines on topics ranging from navigation skills to winter driving.

Introduction

Cumbria's Lake District produces some memorable weather, though unfortunately most of it is memorable for the wrong reasons, as the devastating floods caused by Storm Desmond in 2015 demonstrated only too well.

However, every now and then the grey clouds disperse and we're treated to one of those 'blue sky' days that are so magnificent, so sublime, that one can't help but feel uplifted with a sense of euphoria and joy. It was on such a day in the winter of 2016 that my old climbing pal Wilf Williamson and I set off to ascend Blencathra via the infamous Sharp Edge, a delightfully defined, rocky ridge that provides a strikingly direct line to the summit plateau. An accident black spot, the edge has been the scene of numerous incidents and fatalities over the years, particularly in wet or snowy conditions, so we were mentally prepared for this and appropriately equipped and experienced.

Our walk to the ridge was very pleasant and we reached Scales Tarn, a small body of water nestling in a high corrie beneath Sharp Edge. A steep shale path led up to the start of the edge proper, and we could see that the sun had melted snow from the south side of the ridge, whereas the north side was in the full and determined grip of winter, plastered with ice and rock-hard snow. Though well within our capabilities, we took great care in picking our way along the ridge's craggy crest, staying well away from the treacherous terrain in the shade to our right.

The final steep slopes leading to the summit plateau were streaked with ribbons of hard snow that required care and concentration, but before long our rocky scramble had finished, and we stood on the gentle slopes of the plateau, difficulties behind us, gazing across sparkling snowfields to distant views of dozens of snow-capped peaks etched against a deep blue sky. It was magnificent.

A year or so previously, I'd invested in a pair of ultra-lightweight walking boots that I'd tested in a wide range of conditions. In particular, their grip on slippery surfaces such as snow was very impressive. As we walked over the frozen snowfields, I noticed that Wilf was having to use his boots to kick steps in the hard snow and was finding the going quite tough, whereas I, in my 'super boots', was able to walk comfortably on the icy surface.

I thought I should point out to Wilf how superior my choice of footwear was, but a moment later and before I could utter a word, my feet went from under me in a flash, my arms flapped in a vain attempt to retain balance, but the inevitable happened and I landed very hard, very painfully and very embarrassingly on my coccyx.

I'd fallen (in this case literally) for the oldest trick in the accident prevention book. It's like an inverse take on a magician's sleight of hand that relies on distraction. I was distracted partly by the stunning beauty and the banter of the moment and partly by the focus on one issue, the grip of the boot. Now I've run hundreds of winter mountaineering and winter hillwalking training courses. I've gone to great lengths to help people gain skills and confidence and learn about the hidden dangers that can catch the unwary in hard, wintry conditions. I've taught them to walk and climb on snow and ice, to use crampons and ice axes, to survive and enjoy the unique rewards

Sharp Edge, Blencathra, Lake District.

that come from being on the mountains in deep winter. I've participated in dozens of rescues in similar conditions and even witnessed, many times, people falling in exactly the same way I did. So what happened?

It's best demonstrated by looking back to an injury that befell a member of our mountain rescue team returning from a rescue one dark winter's evening. The rocky path we were descending was frozen hard, and surface water had frozen solid. On top of this was a light covering of new snow, and it wasn't possible to tell if your next step was going to be onto rock with a light cover of snow (OK) or hard ice with a light cover of snow (not OK!). We put crampons on; some didn't. Minutes later I heard an agonised shout next to me. I turned, and etched in the light of my headtorch I saw one of our larger team member's feet shoot out from under him and kick the air in vain before he landed on his back with a thump you could hear in the car park a quarter of a mile away. Our rescue tally for the night suddenly doubled!

That's exactly what happened to me. Though the majority of the summit snowfield consisted of firm snow that my boots gripped supremely, there were a very few small patches of ice, lightly covered with a dusting of snow, that were indistinguishable from the rest of the snowfield. I'd strayed onto one of these, slipped and landed flat on the base of my spine. In my career I'd shown hundreds, if not thousands, of people this very hazard, and I'd witnessed lots of people falling in this way, yet it still happened to me.

Why? As I intimated earlier, it was to do with distraction, deception and perception.

The first factor relates to perception of risk. Our route that day took in Sharp Edge – an accident black spot, some of which was encased in hard snow and ice and from which a fall could be deadly. Our focus was on this part of the route, and having completed it, our perception of risk reduced significantly. This is the deception, though unlike the magician, we are deceiving ourselves rather than an audience. Moving from a very high-risk situation to a lower-risk situation does not mean that risk has been removed, yet this is the deception that we experience.

The second factor is distraction. In this case, I was distracted and lost focus due to two factors – the beauty of the moment and the humour of the moment. The beauty of the moment is important and an underrated factor as a distractor. The scenery and blue skies created a sense of euphoria and happiness that in turn reduced my focus on risk. The humour of the moment, the banter between friends, also acted as a distraction from the potential risks. End result – a slip, a fall, a big bruise and an even bigger laugh for my friend.

I reflected on that momentary lack of concentration and resulting slip and wondered how I could prevent it ever happening again. I came to the conclusion that I couldn't, but what I could do was reduce the chances of it recurring even further, to a level at which it was highly unlikely to happen. To do this, I would need to rethink my whole attitude, develop a different, free-thinking approach to hazards and hone this to a whole new level.

Accidents in the workplace and at leisure cost us individually, cost and inconvenience our families and cost business and society vast sums, yet in recent years the statistics show a gradual flattening in the level of accidents, injuries and time taken off work due to ill health. This is at a time when our knowledge and understanding of health and safety and the levels of support available are at an all-time high. I believe that in order to make further reductions in the numbers of accidents and incidents occurring in the workplace, and to reduce absence due to ill health, we need to look beyond the excellent work that has been done to improve safety through systems, procedures, training and equipment to new ways of managing the human component, new ways of understanding and dealing with factors such as distraction, failure to concentrate or overconfidence.

I'm utterly convinced that by approaching health and safety from an additional, different perspective, and by focusing on reducing human error through the way we look after the health, well-being and fitness of our workforce, it's possible to significantly reduce the numbers of accidents, incidents and near misses and to reduce the number of working days lost due to ill health and injury.

Many factors contribute to the occurrence of accidents, and often it's a combination of factors. Common contributors include distraction, failure to concentrate, not following instructions, forgetfulness, tiredness, changing circumstances, overconfidence and risk normalization. My stance is that by helping people to concentrate better, to remain focused for longer, to feel less tired, to assess risk more effectively, to identify hazards earlier and so on, we can incrementally make a real difference.

To do this, we may have to change the way our workforce eats and drinks and consider whether people are taking their breaks at the best times. We need to ensure their physical and mental well-being are taken into account and improved. Responsibility for health and safety issues needs to be shared, and every person needs to feel involved and listened to. Our workforces need to start thinking differently and adopt a free-thinking approach to hazard identification. We need to manage experience effectively and reduce the overconfidence and numbness to risk that can appear over time. Recognition that people have different learning styles is essential, and we need to cater to this in our training and systems. Increasingly, we may need to consider the demographics of an ageing workforce alongside cultural and ethnic considerations. By accumulating small gains from these and many other ideas described here, we can make a real difference.

A healthy, fit and happy workforce trained in free-thinking hazard identification, knowledgeable about their diet, working in effective teams and genuinely taking responsibility for health and safety will be more effective, more productive and safer.

1 The History of Health and Safety

Health and safety are no different from any other sphere of life in that if we want to move forwards, it's a sound and well-proven idea to take a look at what's already happened.

Although we may trace the official structure of health and safety back perhaps to the early Factories Act of 1833, health and safety have been thought about and acted upon for thousands of years, in spirit if not in name. Amazonian tribes, for example, have hunted with darts tipped with poison as dangerous as most hazardous substances found in a modern chemical factory for thousands of years and they found ways of doing so safely.

In the 1700s – the 1700s BC, that is, the Ancient Babylonians are frequently credited as being the first civilisation to legislate to provide safer working practices, with Law 229 of The Code of Hammurabi:

> If a builder build a house for a man and do not make its construction firm, and the house which he has built collapse and cause the death of the owner of the house, that builder shall be put to death.

Well, it would certainly make you think twice about the quality of your work! This oft-quoted phrase doesn't quite work for me as the start of health and safety, because it's simply punitive and does not address the issue of safety during the construction process.

Q: 'What did we have in place before health and safety?'
A: 'Natural selection'

This old joke does have an element of truth in it, because in earlier times, safety tended to have more of an individual or family-based bias. Many earlier civilisations exhibited some form of collective working, but often it was family based, with the basic need of protecting the individual and family taking priority.

As civilisations developed, larger-scale construction projects became commonplace and these inevitably involved huge amounts of labour. Building enormous pyramids, complex temples, amphitheatres, magnificent cathedrals and great cities, our ancestors worked collectively in roles that often carried high levels of risk. Those at the foot of the labour pyramid would have undoubtedly been exposed to the greatest risk, and as at many times in human history, when human labour was cheap, so too was life itself.

It's estimated that during the building of the Great Wall of China, hundreds of thousands of workers could have died as a result of overwork, malnutrition and the extreme working conditions. In the 1860s, 120,000 workers are estimated to have perished during the construction of the Suez Canal, from a total workforce of 1.5 million – a fatality rate of 8%. During the construction of the Panama Canal, 25,000 people died from tropical disease, falls and crush injuries and the extremely hot and humid conditions. The Karakoram Highway might be the most magnificent stretch of road in the world, but 1300 people died in 1978 during its construction – many being buried in landslides.

Protective equipment would have been almost nonexistent, working hours long and hard, and safety very much left to the individual to deal with.

The development of civilisations and society, coupled with the parallel development of communication, has led to radical shifts in the way that safety has been perceived and dealt with.

In the United Kingdom, the industrial revolution sparked a period during which profit became the driving force of a new industrial society. Profit as a priority, along with an established class system, produced a society in which profit was unashamedly built on a largely uneducated workforce that had a low life expectancy and little hope of progression and personal betterment. The labour force was under great pressure to work long hours in dangerous conditions simply to exist at a basic level. Mines, mills and factories were dangerous places and only our human, innate individual instinct for survival plus the attitudes of some more enlightened and thoughtful bosses prevented even higher levels of injury and mortality.

The Factories Act of 1802 and several subsequent acts prior to 1833 offered a glimmer of hope during an age renowned for its tough working conditions, early mortality and child exploitation, though their collective efforts to impose a maximum 12-hour working day for children in the cotton mills were widely evaded and ignored.

The Factories Act of 1833 introduced the first really serious attempt to combat difficult and dangerous working conditions, child exploitation and excessive working hours. The Act had little support from many quarters and proved very difficult to enforce, with four inspectors having to oversee 3,000 textile mills. Some of the regulations introduced in the Act were visionary and retain significance almost 200 years later. Fitting guards to machinery and the reporting of accidents were two areas covered by this act that are still critical components of modern health and safety.

A generation later, the Factories Act was extended to cover most workplaces, and inspectors were appointed to provide advice and ensure the laws were fully understood. However, the 30 or so people who worked in this role would have struggled to provide this service across the whole country.

1842 saw the introduction of the Mines Act following an investigation into the working conditions of miners. The commission found that accidents, long working hours and seriously dangerous and difficult working conditions, along with maltreatment of workers, were found to be the norm. Public outcry resulted and the Mines Act was brought into force.

The first inspector, Hugh Seymour Tremenheere, had limited powers, but undertook many prosecutions, investigated the condition of the mining community and made recommendations for training managers, reporting of fatal and serious accidents and provision of pithead baths and suitable habitation for mine workers. Later that century, the Quarries Act and the Metalliferous Mines Regulation led to the formation of the Quarries Inspectorate in 1895.

Progress was clearly very slow during the nineteenth century as far as health and safety were concerned, but one has to place this in the context of the time, an age during which massive technological advances were being made and in which engineering and construction underwent unparalleled development. The emphasis was very much on the creation of new designs, the development of new materials – many of which were harmful either in manufacture or as an end product – expansion and profit. The safety of the workforce was, sadly, somewhat left behind.

The first half of the twentieth century saw few changes, in part due to the effects of two world wars, and it wasn't until 1956 and the publishing of the Agriculture (Safety, Health and Welfare Provisions) Act that the safety of workers was once again brought to the fore, this time in Britain's highly important, increasingly mechanised and dangerous agricultural sector. Basic provisions for first aid and hygiene were made alongside setting out requirements for reporting and investigation of accidents and illnesses. The lifting of excessive weight was also controlled – although this didn't prevent my father carrying 100-kg sacks of corn about the farm where I grew up!

This last comment illustrates one of the big problems of that period in that both workers and management were brought up in a world significantly devoid of a health and safety culture. It's difficult to change entrenched human behaviour and expectation, so rather like in the case of the early Factories Act, many people carried on working the same way they always had, and it took many more years, lots more accidents and much more legislation before health and safety culture became more firmly embedded.

1974 saw the implementation of the Health and Safety at Work Act. Employers and employees were consulted during the formation of the Act, following which the Health and Safety Commission and Health and Safety Executive were established, the latter at the start of 1975. The HSE, along with local authorities, is still the major authority when it comes to enforcing health and safety today.

I can look back on this period with a degree of personal experience. At the time, I worked on the farm part time, and undertook some work experience with the local authority, in their forestry and parks and recreation departments. I can say with certainty that at the time I had no knowledge of the 1974 Act, and absolutely nothing changed with regard to the established working practices or my employer's attitude. In short, from a worker's perspective, nothing appeared to change within a large local authority following the publication of the Act. Once again, I'm sure that this was connected with the fact that health and safety culture was still new, not yet embedded in many sectors, and the local authorities found it difficult to implement. Accidents took place, but our collective response in terms of putting in place measures to prevent the same thing happening again was very poor. I recall our chief forester splitting his foot almost clean in two with a felling axe after it slipped sideways off a log he was attempting to split. He didn't wear safety boots, but the accident didn't prompt our employer to ensure that everyone wore them.

That said, the ensuing 30 years saw reductions in fatal and nonfatal injuries at work of approximately 70%, a figure that demonstrates the success of this and many further Acts in establishing a successful health and safety culture. It's also an interesting illustration of the length of time required to change working culture.

One of the subsequent key pieces of legislation was the Notification of Accidents and Dangerous Occurrences Regulations 1980. This required both employers and the self-employed to keep records of accidents and a range of other dangerous occurrences. The Reporting of Injuries, Diseases and Dangerous Occurrences Regulation 1995 – more commonly known as RIDDOR, superseded this Act. This legislation is vitally important not only in providing the basis for statistical analysis of accidents and incidents, but also in providing a continuous flow of information from which crucial lessons can be learned.

It wasn't until 1981 that the Health and Safety (First Aid) Regulations came into force to ensure that adequate first aid provision was available at all workplaces. This included ensuring that first aid equipment was available along with first aid training and the provision of information to employees about where the equipment and trained help was. I find it quite astonishing that it took until 1981 for such a basic provision to be addressed.

The later 1900s saw regulations come into force regarding pesticides, the control of asbestos at work, noise at work, manual handling, display screen equipment, provision and use of work equipment, personal protective equipment, construction (design and management) and construction (health and safety welfare).

It's interesting to look back at these regulations and observe the delays between general and open knowledge about a particular issue and the creation of regulations. For example, the dangers posed by pesticides were well known in the 1960s and 1970s, yet it took until 1986 before regulations on control of pesticides came into force. Asbestos was linked with lung disease in the 1930s, but regulations on control of asbestos at work did not appear until 1987.

On a similar theme, it has often taken a disaster of some type to kickstart health and safety action. The 1987 King's Cross Underground fire killed 31 people and injured many more. The Fennell Inquiry that followed identified many issues, including poorly trained staff who were 'woefully inequipped to meet the emergency that arose'. London Underground accepted 157 recommendations for safety improvements.

The Clapham train crash in 1988, in which 35 people were killed following a triple train collision, led to 93 recommendations for safety improvement and addressed the working hours of signalmen along with the installation of automatic train protection systems.

In 1988, the Piper Alpha offshore oil platform was destroyed following a series of explosions and a gas fire. Out of a total workforce of 226, 165 died, and the Cullen inquiry that followed appointed the Health and Safety Executive as the single regulatory body with responsibility for enforcing health and safety in the offshore oil and gas industries.

The Lyme Bay canoeing tragedy in 1993, in which eight youngsters on a commercially operated canoeing trip died, led to the formation of the Adventure Activities Licensing Authority, which in 2007 was integrated into the HSE.

The reason for mentioning these disasters is to reinforce the point that much health and safety legislation has often been reactive – a response to either a sudden, disastrous incident or the creeping realisation that substances we work with, or working practices, need to be controlled because they present unjustifiably high levels of risk.

This is a diametrically opposed response to the aims of this book, which are primarily to suggest ways of proactively managing the human factor in relation to health and safety.

The early 2000s saw the launch of a 10-year 'Revitalising Health and Safety Strategy' and the 'Securing Health Together' strategy. This was in response to figures suggesting that the same proportion of people had been injured at work since the early 1990s. A further strategy was launched in 2009, aiming to develop 'Renewed momentum to improve Health and Safety performance'. This was in response to figures suggesting that the combined rate of illness and injury was the same as in 2004. These types of strategies highlight a different approach to health and safety at work – a move away from legislation, inspection and a focus on procedures towards a recognition that it was necessary to periodically reassess our approach towards health and safety and focus on illness and well-being alongside accident and injury.

At this point, it's worth a quick look at the HSE statistics. Fatalities at work have continued to decline overall since 1981, with a levelling off over the last 10 years to a figure of approximately 0.4–0.5 deaths per 100,000 workers. The number of injuries has shown a similar decline and a similar leveling-off trend since 2010. Instances of work-related ill health and occupational disease also fell through to 2011, from where they have risen very slightly. In broad terms, our health and safety policies and procedures have been working well, but the figures, not unexpectedly, have tended to plateau in the last few years. However, statistics in a general sense conceal other statistics hidden within them. The figures above cover all types of workplace and if, for example, the number of employees in low-risk occupations rises and the number employed in high-risk occupations lowers, a breakdown in those overall figures could portray a different story within different sectors.

A couple of excellent, though troubling, examples of the statistics hidden within these overall trends is that 25% of fatal accidents involve workers over the age of 60, despite those workers making up only 10% of the workforce, and that agriculture has 8.61 fatalities per 100,000 workers compared with just 1.82 fatalities per 100,000 for construction.

Two further reports were published in 2010 and 2011. The first, Lord Young's review of health and safety – 'Common Sense – Common Safety' – was commissioned by the then Prime Minister David Cameron. 'The aim is to free business from unnecessary burdens and the fear of having to pay out unjustified damages claims and legal fees. Above all it means applying common sense not just to compensation, but to everyday decisions once again'.

The rationale behind the report is not difficult to understand, and in some ways the background to the report exemplifies the process of marginal change. Fifteen years ago I recall my father-in-law, an engineer involved in health and safety training, getting quite hot under the collar at the way in which health and safety were becoming increasingly frequently used as a scapegoat, an excuse for not doing something. He was steadfast in his view that health and safety were first and foremost enabling – enabling people to perform risk-carrying tasks as safely as possible and not providing an excuse as to why something could not be done.

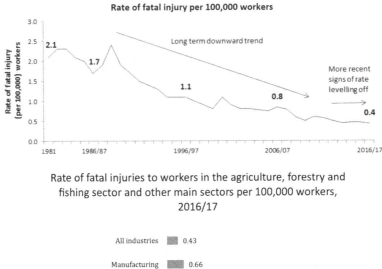

Rate of fatal injuries to workers in the agriculture, forestry and fishing sector and other main sectors per 100,000 workers, 2016/17

The overall statistics can hide unacceptably high fatality rates within certain sectors.

This very negative and limiting side to health and safety grew slowly and steadily, arising predominantly out of a misunderstanding of the principles of health and safety, and perhaps too often out of a misguided individual approach that simply wanted a get-out clause.

A readjustment of attitude towards health and safety was certainly required, and the Young Report addressed a range of issues from tackling 'compensation culture' through to insurance requirements, health and safety requirements in low-hazard workplaces and a reduction in the level of health and safety regulation in terms of school visits and outdoor education.

The second report, 'Reclaiming Health and Safety for All: An Independent Review of Health and Safety Legislation' by Professor Ragnar Lofstedt, was commissioned by the Employment Minister, Chris Grayling. The report was controversial, with many concerns raised by professional health and safety bodies, though many supported the overall aim of streamlining legislation and reducing the burden on British businesses.

Over the years, the health and safety movement has demonstrated significant success in reducing the number of people killed or injured in workplace accidents. As a movement, it gained significant momentum during the latter years of the twentieth century and the early years of the twenty-first century; indeed, some saw it as a self-perpetuating behemoth in need of some sort of restraint. As with many initiatives

New and long-standing work-related ill health per 100,000 workers

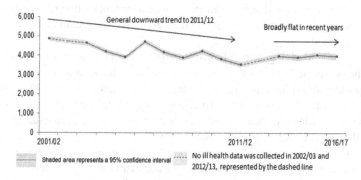

Shaded area represents a 95% confidence interval ···· No ill health data was collected in 2002/03 and 2012/13, represented by the dashed line

Rate of self-reported musculoskeletal disorders
(LFS, England & Wales estimated rate per 100,000 workers)

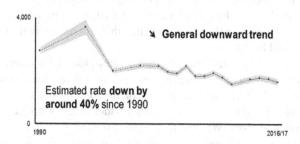

General downward trend

Estimated rate **down by around 40%** since 1990

Rate of self-reported stress and related conditions
(LFS, England & Wales; estimated rate per 100,000 workers)

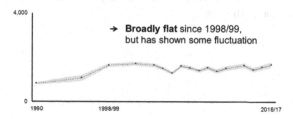

Broadly flat since 1998/99, but has shown some fluctuation

Rate of self-reported workplace non-fatal injury
(LFS: Estimated rate per 100,000 workers)

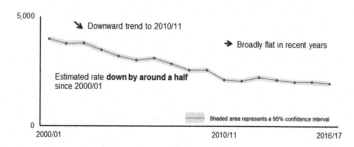

Downward trend to 2010/11

Broadly flat in recent years

Estimated rate **down by around a half** since 2000/01

Shaded area represents a 95% confidence interval

HSE statistics show a general levelling-off trend.

that affect society in such a major way, there is a tendency for the pendulum to swing to either side of an idealistic central point, and recent feeling reflects this, with concerns about some aspects of health and safety being overzealous, too restrictive and overlegislated.

For me, the underlying principle historically is that of finding ways of enabling tasks to be performed safely by reducing the risk to an acceptable level. Equipment design, protective equipment, training, effective systems and procedures all help towards this end, but to progress further and continue to see a fall in the number of accidents, injuries and near misses, we need to have a fundamentally different additional approach that centres on people as well.

There is clear evidence that this approach is now being used in some sectors. Leading the way have been some of the major construction projects such as London 2012, Crossrail, Tideway and HS2. With stated zero-tolerance approaches to accidents and a clear aim of placing safety ahead of all else, these projects have also started to address the human component of health and safety management in a different way, for example, with the introduction of a range of measures that aim to monitor and improve workforce fitness, in the recognition that this will reduce both accident levels and incidences of ill health.

After a period in which accident and injury rates along with levels of ill health in the workplace have stagnated, the time is ripe for a new approach to managing workplace health and safety. We need to build on the successes of some of the largest construction projects and roll them out across a wider range of project and business sizes and organization types. We need to recognize the crucial role that managing the human component of health and safety can play and embed this into every place of work. Over time, this will make a real difference, and as a result we should expect to see a further decline in workplace accidents and injuries and a significant reduction in time taken off due to ill health.

2 A Personal Perspective
Part 1

My working life started prior to the 1974 Health and Safety at Work Act, so like everyone else around my age, I've witnessed the development of health and safety from prior to the inception of the phrase right through to current times, 45 or more years of modern health and safety progression.

I started work in the early 1970s as a schoolboy working on our small, mixed farm down in Yorkshire. We had chickens, cows and pigs and grew potatoes, barley and grass for hay. Surviving this period was one of my personal triumphs, such were the number and seriousness of the hazards farming presented! I don't think the phrase 'health and safety' had been invented at that point, and the thing that stands out most to me is how little thought was given to safety. There was no safety culture, and very few deliberate actions that would have reduced risk, beyond the occasional 'be careful'.

My father had a Fordson Major tractor, a brute of a machine that seemed determined to inflict damage upon anyone who was brave enough to work with it. You needed forearms of steel to cope with the primitive steering, and hitting a bump in the wrong place at the wrong time could easily have resulted in the steering wheel fracturing your wrist.

A long, downhill road led from the farm to a set of fields we farmed a couple of miles away and I thought nothing of sitting on the Fordson mudguard while my father freewheeled down the hill at breakneck speed. Contact between the mudguard and my bottom was merely periodic as the machine bounced and swerved down the road. There was no rollcage or cab and in retrospect, though it was commonplace at the time, it was absolute, undiluted madness.

When I started rock climbing in 1975, technique and equipment were basic to say the least. We used hawser-laid ropes and heavy, steel carabiners and we tied the rope around our waist with a bowline knot. In order to hold a falling climber, we wrapped the rope around our waist and twisted one arm around it to give extra grip. Using physical strength and the friction of the rope around the waist and arm did allow a climber to hold their partner, but the thin rope felt like it was cutting you in two, and severe burn injuries from rope slippage were common.

In the same period, motorcyclists often didn't wear helmets and car seat belts were rarely fastened. Drunk driving was commonplace and there were few hi-vis jackets or helmets on work sites. As a society, we just weren't safety aware; there was no cultural embedding of safety, and employing safety measures was often seen as being soft.

It's interesting looking back at that period, because it illustrates very clearly how a whole culture behaved in relation to safety at work and to leisure and personal life, too. Farmers were, and still are, very self-sufficient, resourceful and also very gung-ho when it comes to getting things done. There was a lot of macho behaviour, risk-taking and arrogance that went along with being a farmer. At the time, that's how it was, and that was the expectation, but if I were presented with a similar set of working conditions now, I'd have a totally different approach, my working life having coincided with the development of modern approaches to health and safety at work and the unstoppable health and safety movement. Agricultural accident figures are still very high as a percentage compared to other industries. Farming has changed, but I don't think farmers have!

It's taken over 40 years and nigh on two generations to create a strong safety culture in the United Kingdom, and it's still developing. Few of the strategies outlined in this book have an immediate impact; in fact, it could take months or even years for their effectiveness to be fully demonstrated. We need to make sure that we're in this for the long haul, and work collectively towards the cultural change that's required for long-term gains.

3 Born to Make Mistakes

John Long is a climbing legend. An American climber, part of the evocative free-climbing movement that dominated rock climbing in spectacular centres such as Yosemite in the 1970s, John Long also developed into an accomplished and entertaining writer, covering everything from climbing instruction to more philosophical writings on climbing. He made a series of groundbreaking first ascents of huge climbs in areas such as Yosemite and was renowned for his skill, strength and mental stamina. If anyone could be given the title 'master of rock', it would probably be him. Yet deep into his climbing career, John Long sustained serious leg injuries in a fall at an indoor climbing wall.

> John Long, who was seriously injured in a fall in a climbing gym, reports that he is on the mend, and has given details about the accident. Long says that he will go in for a third surgery on Saturday to have plates installed in his leg, then will have a final operation to graft the wound that resulted from the break. Long spent a total of 32 days in hospital recovering from a severe compound fracture to his leg.
>
> 'The docs tell me I'll have a 100% recovery,' he says. 'I expect my ankle to be stiff, but considering what could have happened, I am really lucky'.

Long was climbing at the Rockreation gym in Los Angeles on November 29 when he started lowering from an anchor about 30 feet up the wall. 'I went airball, right to the deck,' he said. 'People who saw the fall said I did some juke cat move midair to orient myself feet first. If I hadn't done that,' he says, 'I would have gotten a lot worse than a broken leg'.

Long said that he ties in with a double bowline, but this time, distracted and tired after a long day of work, he didn't finish the knot. 'I made the two bowline loops,' he says, 'and threaded the rope through my harness, but I didn't bring the rabbit out of the hole and around the tree'. Adds Long: 'A lot of people are down on the bowline, but the same thing would have happened with a trace-8. I just wasn't paying attention'.

John Long's message in the last sentence is simple, yet critical – he's saying that lots of people are 'down on the bowline'; in other words, it's not a popular knot – but he goes on to say that it was not the knot that failed, but the process of tying it, during which he wasn't paying attention. He considers the result could have been the same regardless of which knot he was using to tie to his harness.

I've used John Long's accident to illustrate that tiredness and distraction, and indeed all the other common human failings, can affect even the most experienced people.

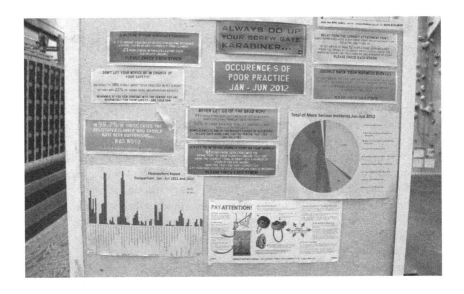

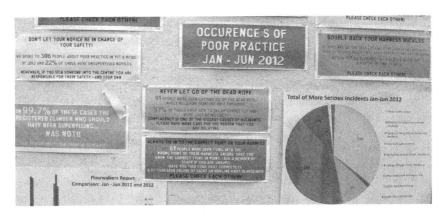

Some climbing walls list their own statistics showing the types of near misses and accidents that have taken place. Climbing walls remain a classic example of how a warm, indoor, nonthreatening environment without objective dangers gives most people a false, low perception of risk.

We're Only Human, Born to Make Mistakes.

Chernobyl, April 1986. A safety test went spectacularly wrong while the operators were simulating a full power failure in which safety systems were deliberately turned off. A combination of design flaws and the reactor operators arranging the core in a different way to the checklist for the test eventually resulted in an uncontrolled reaction. A violent steam explosion and subsequent open-air graphite fire produced radioactive emissions for about nine days. Practically all of this radioactive material went on to fall out/precipitate onto much of the surface of the western USSR and

Europe. Safeguarding and decontamination ultimately involved half a million workers and cost an estimated 18 billion roubles. The effects of this were to be felt for years to come not only in Russia, but even here in my homeland, Cumbria's Lake District.

At the time of the incident, the weather patterns drove radiation-laden clouds across the north of the United Kingdom, and the resulting rainfall – heaviest as always in upland areas – deposited large amounts of radiation on the Lake District hills. As National Park Rangers, we spent a fair bit of our working life out on the mountains, but during this period of rainfall, we were given a strong tip-off – don't go onto the hills. In the years to come, many farmers were unable to sell their stock, as the radiation levels in the animals were too high as a result of grazing irradiated land.

The topic of Chernobyl came up recently in conversation with my youngest son, who knew much more about the subject than I. He was confident that our increased knowledge, our learning from past mistakes and new technology would make nuclear power a safe energy supply for the future. I begged to differ with his youthful optimism, and told him that nuclear power would never be completely safe because of one single factor – human error – our propensity for making mistakes. There'll be another nuclear accident and there will continue to be other environmental disasters such as oil spills and large-scale pollution, because people make mistakes. They always have and always will.

I also put the question to him that how on earth is it possible for a US Navy guided-missile destroyer such as the USS Fitzgerald, with all its sophisticated navigation equipment and highly trained personnel, to collide with a giant container ship in open waters? The USS Fitzgerald accident cost the lives of seven US sailors in June 2017, and early reports state that crew members 'should have spoken up' long before the collision and suggest that the accident was caused by multiple errors by the crew and the failure of two separate navigation teams to take action in the minutes leading up to the collision.

Early in 2018, amid increasing concern regarding North Korea's development of nuclear weapons, Hawaiians received a mobile phone alert from the government stating:

Ballistic missile threat inbound to Hawaii. Seek immediate shelter. This is not a drill.

The warning was repeated on television and radio with the message:

If you are outdoors seek immediate shelter in a building. Remain indoors well away from windows. If you are driving pull safely to the side of the road and seek shelter in a building while laying on the floor. We'll announce when the threat has ended. This is not a drill!

Although the message was corrected 18 minutes later by email, it was 38 minutes later before a text message was relayed to say that a mistake had been made.

Hawaiian State Governor David Ige soon apologized and stated that the message had been sent out due to an employee pressing the wrong button. He went on to say that human error during one of the shift changes was to blame.

Senator Mazie Hirono, a Democrat from Hawaii, tweeted: 'Today's alert was a false alarm. At a time of heightened tensions, we need to make sure all information released to the community is accurate. We need to get to the bottom of what happened and make sure it never happens again.'

Shortly after this, Japanese public broadcaster NHK was forced to issue an on-air apology after issuing an alert incorrectly claiming that North Korea had launched a ballistic missile. The message, received by phone users with the NHK app installed on their devices, read:

> NHK news alert. North Korea likely to have launched missile. The government J alert: evacuate inside the building or underground.

In February 2018, a study commissioned by the Department of Health and Social Care revealed that more than 200 million medication errors are made in the NHS every year. The errors, ranging from giving patients the wrong medication to failing to deliver on time were said to potentially contribute towards 22,000 deaths per year.

Researchers from the Universities of York, Manchester and Sheffield collated the amount of 'preventable errors' in prescribing, dispensing, administering and monitoring medication, and the report stated that the errors could be costing the NHS £1.6 billion per annum.

Health Secretary Jeremy Hunt said that 'part of the change needs also to be cultural, moving from a blame culture to a learning culture so doctors and nurses are supported to be open about mistakes rather than cover them up for fear of losing their job'.

Pharmacists will be given defences if they make accidental errors, rather than being prosecuted, to ensure the NHS learns from mistakes and 'builds a culture of openness and transparency'.

Dr Andrew Iddles, whose 102-year-old mother was given medication intended for the patient in the next bed, said:

> My mother was given medication intended for another patient – a genuine mistake but it could have had fatal consequences.
>
> Whilst we were lucky that the error was recognised and reported, it is encouraging to see the NHS taking positive action to tackle these potentially devastating errors – above all, transparency is essential so staff can admit to mistakes early without being afraid of losing their job.

One could continue with example after example. As humans we are naturally fallible, and we will make mistakes. How we minimize the impact of this fallibility by managing both our actions and the effects of our actions is key.

The Swedes have for some time been implementing their Vision Zero policy in relation to roads and traffic. The Vision's key statement is that no loss of life is acceptable, and the policy works towards a situation in which no lives are lost in traffic accidents. Crucially, the policy not only acknowledges, but is based around, the fact that humans make mistakes. As humans we will be distracted, we may be overconfident, we will make errors of judgment and from time to time we'll make silly mistakes. Their philosophy is that by accepting this fact, we can then build an entire system around it, taking into account, for example, the design of roads, the design of cars, the relationship between designated space for cars and people and the education and training of drivers. Just as the Swedes are trying to build a whole transport system around human fallibility, we need to build our whole health and safety system around the same basic precept and develop a combination of responses that:

- Reduce the number of incidences of human error that occur in the workplace
- Reduce the impact that errors have

There are two key ways of accomplishing this. First, we can provide a strong structural approach through training, equipment, procedures, systems, policies and supervision. Many organisations now provide exemplary and highly successful strategies in this respect, and some of our biggest recent construction projects such as building the infrastructure for the London Olympics in 2012, Crossrail and the Tideway project exemplify this approach.

Second, we can seek to manage the human component outside of the structural approach. Focusing too strongly on a structural approach can lead to us neglect the management of the physical, emotional and psychological factors that can lead to human error and subsequent accidents, incidents and near misses.

This book primarily concerns reducing the number of instances of human error that occur in the workplace through an aggregated marginal gains approach that builds on the successful management of the physical, emotional and psychological needs of a workforce. Many of the examples I have used stem from the world of outdoor adventure, extreme sports and modern, high-level sport.

The key elements of this approach take into account:

- Mental well-being
- Physical well-being
- Injury avoidance and management
- Nutrition
- Sharing responsibility and learning from mistakes
- Managing experience
- Adopting a marginal gains approach
- Free-thinking hazard identification
- Managing changing circumstances
- Working culture
- Risk perception
- Team building

4 Marginal Gain

The use of aggregated marginal gain in sport was epitomised and widely promoted by the approach of the former performance director of British Cycling and Team Sky, David Brailsford, who believed that tiny improvements in every aspect of the cycling process would lead to a significant overall gain. The concept is simple – you might also refer to it as attention to detail – and has been understood, though not applied as rigorously, for years. In sports such as cycling and downhill skiing, where the difference between success and failure can frequently be measured in a few hundredths of a second, it's obvious how very small increases in performance can have a major impact on results.

Any process can be broken down into component parts. Very small, marginal gains in each of the component parts can create a cumulatively large and worthwhile overall gain. It's a very simple concept, though one should not underestimate the imagination, thoroughness and intensity of thought that allowed the British Cycling Team to take full advantage of it.

Not everyone believes in the theory of marginal gains – Sir Bradley Wiggins is reputed to have called it 'a load of rubbish', stressing the point that physical ability and training were the real keys to success. However, Brailsford believed that the successful execution of this strategy would lead to Team Sky winning the Tour de France, which they did courtesy of Sir Bradley Wiggins in 2012. In the same year, the British cycling team dominated the London Olympics. Something was working!

In mountaineering, there have been several periods in which the theory of marginal gains predated Brailsford and has been implemented with positive results.

The first is all about weight. Climbing a mountain or hiking a long-distance trail feels easier and is faster if you're carrying less weight. Thirty or 40 years ago British climbers (with exceptions) were well known for being rather slow on big Alpine climbs compared to their European counterparts. Many of them carried bivouac equipment in case they needed to sleep out overnight, along with extra food and clothing. Of course, carrying all this extra equipment slowed them down, sometimes to the point where they needed to use them. 'Don't take it and you won't need to use it' became the mantra for a new breed of British alpinist in the 1980s. Their approach led to faster and harder ascents and inspired a generation of climbers to expand their lightweight philosophy into the world's bigger mountain ranges.

With care and the right levels of experience, this lightweight approach works well and enables mountaineers to climb faster, spending less time in the danger zone exposed to objective dangers such as stonefall and avalanche, and enabling climbs to be completed faster. In this respect, speed and efficiency equal safety.

Equipment manufacturers worked and continue to work on making equipment lighter, and mountaineers worked on reducing the amount of equipment they carried. Carabiners, used to link climber to rope and safety equipment, provide a prime example. In the early days of harder rock climbing – the 1960s and 70s – carabiners commonly weighed about 100 g each. Now, you can buy a vast range of carabiners with weights of between 50 g and 25 g each. Bearing in mind that a modern climber on a reasonably long climb will carry about 40 carabiners, the weight saving starts to make a real difference, with the total weight in the examples quoted above being 4 kg, 2 kg and 1 kg, respectively. Try hanging 4 kg, then 1 kg, from your climbing harness and believe me, you'll feel the difference! This principle of weight reduction has been applied to all climbing equipment to some extent, resulting in an overall equipment weight reduction of probably 50%–60% compared to just 20 years ago – a prime example of how using an accumulation of marginal gains can have a major impact. On the very hardest rock climbs, so tenuous and precarious is the climber's grip on the rock face that every gram makes a difference.

The Duke of Edinburgh Award Scheme has enabled hundreds of thousands of young people in the United Kingdom to experience adventure, teamwork and the benefits of community projects. It's also allowed them to gain useful outdoor skills and enjoy walking and camping in some of our most beautiful areas. The Gold Award challenges young people to carry out a multiday expedition in 'wild country' – often in areas such as the Lake District, Snowdonia or Dartmoor.

Over the years, I've acted as assessor on a number of these expeditions and in almost every case, the expedition members carried way too much with them, sometimes to the point where they could barely lift the juggernaut rucksacks onto their backs. Truly the HGVs of the outdoor world! I was always surprised at what was packed away in their bags. Individual wash kits, for example, that contained enough to last the whole group a week or more; heavy tins of food; large hand-held torches; boxes of washing powder – I never saw the kitchen sink but it sometimes seemed likely to appear in the bottom of the bag!

My wife had a telling experience while she was undertaking her teaching course, which had a strong outdoor education component. One of the outdoor instructors for the course was very intense and rather traditional in approach. On a pre-expedition check, the instructor was systematically going through everyone's equipment and when she came to my wife's bag, she lifted it up and said that it couldn't possibly contain everything that was required because it was too light! Out came the contents of the bag to reveal why it was so light – every item had been rationalised and reduced to its minimum state – a perfect example of marginal gains. The toothbrush handle was cut in half, and just enough toothpaste had been decanted into a tiny plastic bag. One spork replaced the traditional knife, fork and spoon, and so it went on, each gram of weight saved combining to provide a significant overall reduction. At the end of the inspection, the instructor had to reluctantly agree that all required items were present. Their attitude was that heavy weight of bag = safety. My wife's attitude was diametrically opposed – light bag = safety.

The reality is that carrying heavy loads on mountain expeditions is often unavoidable, but the heavier the load, the more energy you expend and the more difficult it is to move over challenging terrain. In the case of D of E expeditions, the weight of the packs has undoubtedly been a significant factor in many instances of exhaustion or exhaustion/exposure (a combination of fatigue and hypothermia) and in some instances of injury due to falls and slips.

Another example of marginal gains from the rock-climbing world concerns speed rather than weight. On longer, roped climbs, the lead climber stops periodically to set up an anchor and bring the second climber up to the same point. Each section is called a 'pitch', and at the end of each pitch, the lead climber secures him- or herself to the rock, the 'belay'. On long climbs, there could be 10 or more pitches, requiring the same number of belays to be set up.

Each time the lead climber reaches the end of a pitch, he or she has to do a number of things, in a certain order. First, he or she has to secure him- or herself to the rock face by either attaching him- or herself to fixed and permanent anchors, or by placing his or her own anchors and attaching to them. Once secured, he or she pulls in any slack rope that remains between him- or herself and the second climber, then attaches a belay device to the rope that enables him or her to hold the second climber should he or she fall. The second climber ascends to join the first, while the lead climber keeps the rope taut between them by taking the slack rope in through the belay device. Once both climbers are together, they have to sort the equipment out for the next pitch, rearrange the ropes and belay to suit, and the lead climber then sets off again.

There is therefore a lot to do while no actual climbing is taking place. If, at the end of each pitch, it takes the lead climber 10 minutes to set up the belay and prepare for the second to climb, you can see how over 10 pitches, it would take up to over an hour and a half. Over 15 pitches, it would take two and a half hours. If, by a combination of increased efficiency, skills training and experience one can cut the time taken to set up each belay to 5 minutes, over 15 pitches the time saved would be one and a quarter hours.

That's one and a quarter hours less time exposed to the potential risks of stone fall or the onset of bad weather, and it can mean the difference between getting safely down from the climb in time for tea or getting caught in darkness and exposing yourself to an overnight bivouac or difficult torchlight descent.

To achieve such a reduction in time means applying the principle of marginal gains to each of the subcomponents of setting up the belay and rearranging the ropes. The best climbers do this and achieve faster ascents that show no compromise towards the safety of the technical component of the climb, but significantly reduce time spent in areas of objective danger.

A good workplace example to complement the outdoor activity examples would be to consider how one could apply marginal gain theory to method statements. Feedback from several figures involved in major construction projects is that method statements have in some cases become excessively big and wordy, to the point where they have become more of an exercise in box ticking, simply due to the time required to read and understand them, at the expense of acquiring a deep understanding of the planned approach. A marginal gains approach could reduce the size and complexity of all method statements so more time is spent actually understanding them and the risks, rather than just going through the process. Breaking down method statements and looking for alternative forms of communication such as visualisation, demonstration and the use of a wider range of graphic representation such as iPads could increase the ratio between admin time and actual transfer of understanding of approach and risk.

And now for the important question: What can we learn from Brailsford's approach and from the application of marginal gains theory to health and safety?

First, we can apply the theory to the reduction of human error in an overall sense. By making simultaneous marginal gains through the key principles highlighted in this book, it's possible to achieve a significant gain. For example, not everyone will buy into or act upon the theory of eating a specific diet to help maintain a higher level of alertness and concentration. Nor will everyone want to adopt a training regimen to help combat injury through repetitive action or perform warm-up exercises prior to starting work. One hundred percent or even 50% take-up is utterly unrealistic. However, if 30% of a workforce act upon some of the recommendations contained here, we can expect a significant reduction in work-related accidents and injury.

Second, we can learn from the way that marginal gain theory breaks down tasks into component parts, and then looks deeper and sideways for further ways of making improvements. This links strongly with the principle of free-thinking hazard identification, in which a workforce needs to adopt a more thorough, detailed and penetrative approach to identifying hazards.

The application of marginal gain theory in health and safety requires us to consider a number of factors in order for it to be effective.

APPROACH AND UNDERSTANDING

First of all, it's essential that the way in which we approach marginal gains is structured, positive, understood clearly and verbalised effectively. Marginal gain isn't about what might happen or what results we hope to achieve, it's about what we expect to achieve and what we will achieve. Be bold and set high expectations. Managers need to set the process in motion in the right way, using positive, 'will do' language. It's essential that everyone understand what you are trying to achieve and that you are setting outcomes that everyone buys into. Finally, absolute clarity is essential concerning roles, responsibilities, methods and strategy.

ENTHUSING

It's easy to invest time and effort into schemes that a workforce simply won't buy into, but it's absolutely essential that they buy into the marginal gains approach at all levels within an organization. To do this requires enthusiasm for the concept and a great sales pitch to the workforce. You might consider approaching this with an in-house marketing strategy, maybe in conjunction with reward schemes and recognition. Linking with established and well-known sport principles such as cycling, F1 or ski racing might help. As a manager, if you are not fully and demonstrably behind the concept, there's no way that your workforce will comply – you've got to set an example. Celebrate successes when they occur and use success as a lever to embed and develop the marginal gain principles. Half-hearted attempts to improve through marginal gains don't work and can have a negative impact on some people, who may feel disenfranchised and that it's a waste of time. It's essential to make marginal gains a team effort and to take collective responsibility and enjoy collective success.

FEEDBACK

A strong feedback partnership is critical to the success of a marginal gains approach to managing health and safety. I've mentioned elsewhere that in modern society, many people look for the big ideas, the game-changing concepts that provide instant results and recognition. A marginal gains approach relies on a diametrically opposed attitude that an accumulation of small improvements will lead to a longer-term, measurable and significant improvement. Accurate, frequent, shared feedback that is acted upon quickly and efficiently is essential to the success of a marginal gains approach. Management must provide the right example and the right culture for feedback to work effectively.

One of the themes that came back from the questionnaire put out to the construction industry was that although major incidents are reported well, there is still a reluctance for a workforce to report minor incidents or near misses, with it not making any difference and the paperwork and effort required to make the report frequently cited as reasons why it's not done.

Reasons for not reporting incidents and near misses

- Won't make a difference

- Paperwork takes too much time

- Don't want to get into trouble

- Think it's not important

It's important to employ a range of strategies to ensure that all accidents, incidents and near misses are reported.

Management must ensure that there are alternative strategies in place for providing feedback on minor incidents and near misses. Ideas might include a sticker board for people to post notes, or regular, short one-to-one chats. People could also provide feedback through social media or texts; your approach, though, must take into account the age differences in a workforce and the different expectations in terms of communication. Make it easy to provide feedback, make sure the feedback is acted upon and make sure the person who fed back is kept informed as to how the feedback is used. Individuals don't necessarily want to be rewarded, but recognition and a feeling of appreciation are critical.

Another idea to increase and enhance feedback is to give it added importance by sharing it with a wider audience. This could be through collaboration with other departments, with other partner organisations or companies under the same ownership, or through a wider scheme of sharing health and safety knowledge.

INTERROGATION

In order to learn effectively from a marginal gains approach, it's critical that we don't simply respond to incidents or near misses, but that we actively seek ways of questioning working methods to look for marginal improvements. Some individuals do this as a matter of course; it's in their nature. Others don't, and we need to provide structured opportunities for everyone to critically examine their working methods with the aim of finding improvements that will impact positively on health and safety. I'd call this interrogation, as it's more than simply questioning a process, it's a deeper and more thorough approach that will often require facilitation and dedicated time. Dig deep and encourage lateral thinking.

A great example of this is work being undertaken at Hexham hospital, where one of the orthopaedic surgeons has applied marginal gain theory to knee replacement procedures. Questioning, then making improvements to all aspects of the process, from patient preparation through to anaesthesia and nutrition immediately post-op, has made a significant reduction in hospitalisation time, freeing beds faster and aiding patient recovery.

THINK SMALL

From a management perspective, there is another key learning point here concerning our tendency to perhaps overestimate the value of larger, easily measureable and quantifiable improvements, at the expense of underestimating the value of creeping improvement. We naturally prefer to see larger and more transparent outcomes and we frequently put pressure on ourselves to deliver them alongside the knowledge that these sorts of outcomes are more easily recognised and rewarded. Marginal gains take longer to aggregate into measurable outcomes, but the long-term results can be equally meaningful.

DEVELOPMENT

Develop and reinvigorate your programme of marginal gain regularly. Assign clear tasks relevant to making gains to specific personnel – for example, you could try delegating the task of making marginal gains through technology to younger members of a workforce, who might have less overall work experience, but considerably more experience of technology than their peers. Challenge people to come up with development ideas; make it fun and interesting!

MARGINAL LOSS

Managers need to consider that alongside being aware of the longer-term benefit of marginal gains, it's critical to be aware of the reverse effect – marginal loss – in which small, detrimental changes eventually combine to create a problem. One example of this from a health and safety perspective is the tendency of a workforce to start a project with the best intentions regarding safe working, but as time goes on, those short cuts in working practices and the tiny changes that familiarity brings, such as increased complacency, can accumulate and lead to the occurrence of accidents or incidents.

Change tends to be initiated when we realize that something has gone badly wrong. It's often too late and usually only in response to a problem that is large enough for it to already have done some damage. This is to some extent only natural as a behaviour. We like to see hard evidence of negative change before we act. Most people don't want to be seen to be acting hastily, perhaps afraid that their actions will not have the desired effect, perhaps preferring to monitor a situation longer to see if things improve.

When marginal loss occurs, we tend not to see the results until the accumulation of losses reaches a point at which we realize there's a problem, by which time it's too late.

It's therefore essential for managers to be acutely aware that marginal losses in terms of working practices in relation to health and safety need to be measured and reacted to in an appropriate way, as waiting until a change-precipitating accident occurs is simply too late.

5 Accidents, Acts of God and Human Error

TERMINOLOGY

It's not a bad idea for us to be absolutely clear about the terms we are using when describing accidents, incidents, near misses and so on.

An accident is an event that leads to one or more physical injuries. Any event that causes an injury may be considered an accident regardless of the severity.

An accident is also an event that's not deliberately caused and which is not inevitable.

An incident is described as an event that occurs without causing injury. Examples could include a fall, slip or trip that has no physical consequences.

A near miss is a situation in which no specific event occurs, although the situation could potentially have resulted in an incident or accident – "Close call!" Examples could include being temporarily lost in the mountains, but then finding your way again, or realizing at the last minute that you hadn't clipped onto the safety sling on the cherry picker before ascending.

I walked away uninjured from this motorway crash – accident, incident or near miss?

An act of God is an unforeseen natural phenomenon that is not possible to realistically guard against, that involves no human actions and that is due directly and exclusively to natural causes.

A mistake is an act or judgment that is misguided or wrong.

It also refers to something that is not correct, especially a word, figure or fact – an inaccuracy. The word *mistake* comes from the old Norse word 'mistaka', which means taken in error.

An error is rather unhelpfully defined as a mistake, though it can also be a measure of the estimated difference between the observed or calculated value of a quantity and its true value.

Human error can be defined in different ways – for example, here are some common definitions:

> Human error occurs when someone makes a mistake that causes an accident or something bad to happen. (I would amend this definition to indicate someone can make a mistake that does not result in either an accident or something bad happening.)
>
> It's the making of an error as an inevitable or natural result of being human: the making of an error by a person, especially as contrasted with a mechanical or electronic malfunction.
>
> So there you have it – an error is the same as a mistake in the context we're using, and human error is really just another way of saying that we've made a mistake.

HEALTH AND SAFETY EXECUTIVE DEFINITIONS

The HSE has applied a range of definitions and meaning to human failure – starting with two key types of failure:

- Human error – defined as an unintentional action or decision
- Violation – defined as a situation in which someone has deliberately done the wrong thing

The HSE defines three types of human error categories: slips and lapses (skill-based errors), and mistakes. According to the HSE, these types of human errors can happen to even the most experienced and well-trained person.

Human Failure Types

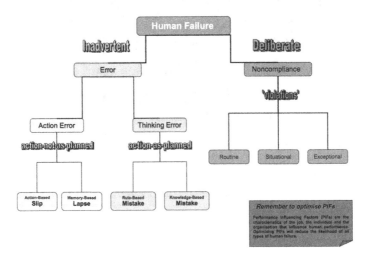

The HSE classification of human error.

SLIPS AND LAPSES

Slips and lapses occur in very familiar tasks that we can carry out without much conscious attention, for example, driving a vehicle. These tasks are very vulnerable to slips and lapses when our attention is diverted even for a moment.

Slips are effectively not doing what you're meant to do.

Examples of slips include:

Misreading a measurement by a point value and subsequently administering an incorrect dosage

Omitting placing axle stands under a jacked-up vehicle

A lapse is defined as forgetting to do something or losing your place midway through a task.

Examples of lapses include:

Forgetting to replenish a saline drip

Tightening, but not torqueing, the wheel nuts after changing a wheel

Mistakes are defined as decision-making failures. We do the wrong thing, but we believe it to be correct. The two main types of mistake are rule-based mistakes and knowledge-based mistakes.

A rule-based mistake could be a misapplication of a good set of procedures or the application of a bad rule.

Knowledge-based mistakes occur as a result of a person having insufficient knowledge or experience to carry out a procedure or make a correct decision.

VIOLATIONS

These are intentional failures, circumstances in which workers deliberately do the wrong thing. The violation of health and safety rules or procedures is one of the biggest causes of accidents and injuries at work.

The HSE identifies:

Routine violations, where noncompliance becomes the norm, perhaps due to a lack of enforcement or a creeping move away from a norm, which develops into a new norm

Situational violations, in which one or more situation-specific factors contrive to make noncompliance the only solution to an otherwise difficult or impossible task

Exceptional violations, in which a calculated risk of noncompliance is taken by a person wanting to complete a task in highly unusual circumstances

I've produced an alternative classification of human error, looking at the underpinning reasons why the error occurs.

HUMAN ERROR CLASSIFICATION HEADINGS

1. Physiological/Psychological
2. Organisational/Personal
3. Deliberate

PHYSIOLOGICAL/PSYCHOLOGICAL

This category includes headings:

Tiredness
Lack of Concentration/Focus
Overconfidence/Familiarity
Distraction
Forgetfulness

ORGANISATIONAL/PERSONAL

This category includes the headings:

Lack of Judgment/Knowledge
Changing Circumstances
Inaccurate Risk Perception
Deadlines
Workload
Confusion
Not Taking Responsibility

DELIBERATE

This category includes the headings:

Clear Aim of Positive Outcome
No Aim of Positive Outcome but without Malice
Intent to Cause Damage/Harm/Disruption

Physiological/Psychological				
Tiredness	**Lack of Focus or Concentration**	**Overconfidence & Familiarity**	**Distraction**	**Forgetfulness**
Physical	*Physical*	Personal nature	*Internal*	Information
Overworking	Physical tiredness	Experience related	Talking or	overload
Long hours	Pain due to injury	Peer pressure	socialising at	Lack of personal
Insufficient	or illness	Repeated task	work	organizational
breaks		many times		skills

Continued

Physiological/Psychological				
Tiredness	**Lack of Focus or Concentration**	**Overconfidence & Familiarity**	**Distraction**	**Forgetfulness**
Muscular fatigue	Lack of sleep	High experience levels with similar tasks	Phones/social media	Medical issues
Chronic injury	Too little exercise		Local work or environmental event that distracts	Distraction, shift of focus
Lack of exercise	*Mental*			Interruption
Medical problems	Depression, anxiety, stress		Interruption	Lack of supervision
Lack of sleep	Phone/social media		Trying to do too many things at the same time	
Dietary	*Local*		*External*	
Anaemia	Attention drawn off-task due to another event or incident		Personal problems	
Inappropriate diet			Relationships	
Caffeine/ sugar-related blood sugar fluctuations	Background noise		Finance	
	Local environmental conditions		Health	
Dehydration	Lack of supervision			
Mental	Interruption interferes with concentration			
Out-of-work relationship, financial or other issues	Boring, repetitive task			
Workplace problems	Work done from habit			
Mental fatigue	Quality of information			
Overexposure to mobile devices or computer	*Nutrition/Medical*			
Diet/nutrition (e.g., too much sugar, insufficient vitamins)	Inappropriate diet			
	Hunger or dehydration			
	Medical conditions (e.g., pregnancy, menopause, anaemia and many more)			
	Medication			
	Drug or alcohol misuse			

Organisational/Personal				
Lack of Judgment/ Knowledge Not Taking Responsibility	**Changing Circumstances & Not Taking Responsibility**	**Inaccurate Risk Perception**	**Deadlines & Workload**	**Confusion**
Not having the required experience, knowledge or training	Environmental conditions	Risk normalization	Self-imposed deadline	Lack of clear instructions
Inappropriate task for that person	Operational conditions	Poor judgment	Organisational imposition of deadline	Not understanding instructions, e.g., due to lack of training or knowledge
Judgment interfered with by third parties/work colleagues	Personal situation	Peer pressure	Too much work expected to be undertaken in given time	Task is too complex
Changing circumstances affect judgment	Unclear as to whose responsibility it is	Not understanding the task and associated hazards	Workload extended due to overtime	Different instructions issued
Organisational failures, e.g., lack of support and monitoring	Lack of supervision		Intensity of workload too high due to lack of breaks	Uncertainty due to changing circumstances
	Personal failing		Not enough work to fill given time	Confusion between similar tasks
	Lack of clear instructions			Conflicting attitudes towards work/tasks and health and safety
	Culture of leaving it to others			Organisational failures, e.g., lack of supervision/ monitoring
	Being careless			Confusion due to mental or physical tiredness
				Psychological factors

Deliberate		
Deliberate Error with Clear Aim of Positive Outcome	**Deliberate, without Clear Aim of Positive Outcome, with No Thought for Consequences, but without Malice**	**Deliberate with Intent to Cause Harm or Damage**
Aiming to make things faster or more efficient	Taking a shortcut within a task (cutting corners)	Sabotage
Aiming to make things safer	Not thinking clearly about results of actions	Actions aim to harm another individual
Aiming to make things easier	Wanting to make life easy	Actions aim to harm the business, property, machinery, etc.
Trying to help colleague with good intent	Following bad example set by others	
Trying to solve a problem in an emergency	Time constraint leads to missing out parts of process	

It's clear that many accidents, incidents and near misses are due to a combination of factors, and that contributory errors on the part of more than one individual are also commonplace. It should be possible to attribute factors from the categories listed above to any accident alongside other contributing factors such as equipment failure.

Below, I've taken some examples from accidents and incidents I've mentioned in this book and used my own classification to attribute cause.

1. John Long, one of the most experienced climbers in the world, falls from the top of an indoor climb and sustains serious leg injuries after failing to complete the knot that ties the rope to his harness.
   Cause:
   Tiredness – Long working hours, possible other contributing factors
   Distraction – Internal – Local environmental factors
   Forgetting to Do Something – Failure to complete knot due to distraction
2. Young person on an outdoor activity trip dies following a jump into a plunge pool.
   Cause:
   Lack of Judgment or Knowledge – Supervisor did not have required experience, knowledge or training; inappropriate task for that person; changing circumstances affect judgment; organisational failures
   Changing Circumstances – Environmental and operating conditions
   Inaccurate Risk Perception – Poor judgment, failure to understand task and associated hazards
3. While out walking, I slipped on ice that was hidden beneath a light layer of snow.
   Cause:
   Lack of Concentration – Attention drawn off task
   Distraction – Internal – Talking/socialising, local/environmental factor that distracts
   Inaccurate Risk Perception – Risk normalization
4. Gustav Fischnaller is badly injured after failing to follow correct emergency procedures to deflate an Upski ski sail, whose release system has jammed.
   Cause:
   Lack of Judgment/Knowledge – Not having required experience, knowledge or training
   Changing Circumstances – Environmental conditions
   Overconfidence/Overfamiliarity – High experience levels with similar tasks (nontransferable skills)
   Inaccurate Risk Perception – Not understanding task and associated hazards
   Confusion – Confusion between similar tasks

It's clear when one examines the causes of different accidents that some of the contributory factors are directly related to procedural issues, while other factors are linked more strongly to human error.

For example, take my slip in the snow. The key contributory factors were all due to human error – distraction and lack of concentration. These errors can be corrected or minimized through free-thinking hazard identification strategies.

Compare this with the accident involving the plunge pool activity, for which the chief contributory factors were a lack of judgment, knowledge and training; organizational failures; changing circumstances and a failure to understand the task and its associated hazards. These errors relate more to training and procedural failures and to organizational inadequacy and a broken system.

6 Lightning Always Strikes Twice

During my early years as a National Park Ranger in the Lake District, I joined the local mountain rescue team, partly as mountain and water safety were an integral part of my job, and partly out of a strong desire to help people who had come to grief in some way while out walking or climbing in the mountains. Up until one remarkable day in the summer of 1991, I'd never been first on the scene at a fatality. That was all about to change.

I recall the weather on that summer day very well. Hot and humid even during the early morning; as the day progressed, the heat built progressively and mercilessly and the humidity rose, developing into one of those sultry, classically English thunderstorm days. The weather forecasts had predicted storms during the afternoon and it looked like they were about to be proved correct.

I'd arranged to visit the Powell family, who farmed the idyllic Easedale valley, at Brimmer Head to discuss some footpath repair work. Over the years, I'd helped the family out with path repairs, waymarking, new stiles, rescuing sheep from crags and so on, in turn developing a relationship that extended to cold homemade lemonade in the farmhouse kitchen, cakes and even the odd wiry chicken thrown in for good measure.

By the time I was driving down the track to the farm that day, the humidity was at tropical levels and the sky darkened by the minute. The first peals of thunder rumbled and the first few huge raindrops splattered onto the dusty ground as I knocked on the kitchen door, to be welcomed as usual by a snarling terrier and Mrs Powell, who ushered me in out of the rain. Even before cups of tea had been made, the storm was upon us. The old-fashioned telephone bell in the house rang repeatedly in time with the lightning strikes and the thunder sounded direct from Thor's hammer. The rain was monsoon in intensity. The electricity went off. Daylight gave way to twilight. After 20 minutes or so, the storm started to ease, but a sharp and insistent rapping on the door abruptly shattered the quiet of the ancient farmhouse.

'Someone's been struck by lightning on the path to the Tarn – can I use your phone to call for help?' The phone was out of action, but my radio was still working, so I contacted one of the other rangers, who initiated a mountain rescue call out via the police. Having made sure that the informant stayed put at the house to direct the rescue team to the location of the accident, I set off at a run across the meadow with a first-aid kit and radio to see what help I might be able to offer and to provide an update to my rescue team colleagues.

After a few minutes jogging up the rocky track, I came to a small group of people huddled over the body of a man on the side of the path. After quickly introducing myself, it became equally quickly clear that the man was dead. I gathered information from the bystanders and built up a picture of what had happened.

The dead gentleman and his son had been fishing up in Easedale Tarn, but as the storm approached, they decided to call it a day and descend. The storm hit with an unusual intensity, and most walkers in the area had stopped to don waterproofs and try to take whatever shelter they could. The gentleman in question had also stopped to take shelter, but sadly decided to sit beneath one of the few trees in the area, a lofty larch that had somehow managed to grow proud and tall from the rocky ground. Sat beneath the tree with his back to the trunk, he held his fishing rod upright and was awaiting the abatement of the storm when the lightning struck the tree with savage force. Each ring on his fishing rod had severed like a broken fuse. His anorak was shredded into small strips and his chest and feet blackened and charred. It was a sobering sight. The rescue team arrived in force a short time later to retrieve the body and draw a conclusion to our role in this sad story.

I recall at the time feeling desperately sorry for the family, but alongside that I was left with a feeling that the man's death had been so easily avoidable – perhaps that was the real tragedy.

There are two very simple and key actions that would have resulted in a different outcome. They both relate to knowledge and the subsequent implementation of actions taken as a result of that knowledge.

First, a look at the weather forecast would have alerted anyone walking in the area on that day to the likelihood of violent storms breaking out in the afternoon. In terms of planning a day out walking and fishing, it would have been clear that making an early start and taking an early finish would have probably resulted in missing the storm altogether. Interestingly, this key point of not reacting to available information is a very common one. I've been on rescue operations many times when the people being rescued knew the reason for their misfortune in advance, yet failed to act in a preventative way.

Second, it is common knowledge that during a storm the last thing that you should do is take shelter beneath a tree – especially in an area where there are relatively few trees – as lightning will tend to strike the highest point in a given area, and that is often a tree.

Given that an average person (a) has access to a weather forecast and (b) probably knows of the danger of sheltering beneath a tree during a storm, I have found myself asking the same questions over and over – why did the man concerned not heed the weather forecast and take appropriate action to avoid the storm, and why did he then proceed to sit under a tree when he was caught in the middle of it?

It is, of course, possible that he did not look at the weather forecast, and it's possible that he did not understand the danger that he exposed himself to by sitting beneath the tree. I doubt that this is the case, though, and I believe the likely scenario is as follows, particularly as there are several themes that are common with many mountain rescue incidents:

1. *Inability to break out of routine*: By this, I'm referring to the incidence of the lightning strike to the routine of a person on holiday, but it could equally be the routine of someone at work. Many people are incredibly inflexible. If the B&B serves breakfast between 8 and 9 a.m., that's when they will have breakfast. The result is that everything else that happens on that day is affected by that timing. To avoid being caught in the storm and therefore avoid being struck by lightning, that routine would have to be broken, and the vast majority of people will not break this sort of routine. Why? Perhaps not wanting to be a nuisance to the B&B, not wanting to miss the breakfast you've paid for or simply not feeling able to move away from the norm and be different.

2. *Ignoring the obvious*: When something is clearly a bad idea, we would all like to think that we would avoid it. However, many people ignore the obvious, cast aside clear warnings and continue regardless. A man I knew told me he lost his life savings through meeting a couple of men in a local pub. They gained his friendship and slowly drew him into their web of deceit. They offered him a chance to invest a small amount of cash with the promise of a quick profit, and he accepted and was delighted when the profit and a bit more besides materialised a week later. They offered him another chance to invest, then another, and each time their promise of a profit came true. The bait was taken, and when the man was offered a chance to make some serious money by investing a bigger amount, he agreed and invested his life savings. He never saw the two men in the pub again.

I spoke to him at some length afterwards, and the end result was extremely sad – he lost his marriage, his B&B and his dignity alongside his cash. There was one burning question, though, that hung in the air throughout our chat, and he eventually answered it without the need for me to ask it. Yes, he'd smelled the warning signs, he'd chewed on the possibility that it was a fraud and he knew deep down that the bait was rotten – yet he swallowed it hook, line and sinker. He saw the obvious warning signs flashing red in front of him, yet chose to ignore them.

It's probable that the man struck by lightning knew it was a bad idea to sit beneath a tree, but he went ahead and did it anyway, ignoring the danger signs and paying the ultimate price.

This is perhaps a good time to introduce the concept of the Ease of Avoidability Axis. This simple method of assessment plots the seriousness of the outcome of the accident against a measure of how easy it would have been to avoid it.

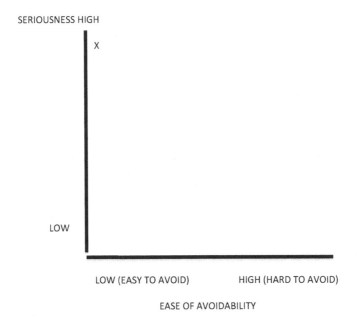

X = POSITION OF LIGHTNING STRIKE ACCIDENT – VERY SERIOUS (FATAL) BUT EASY TO AVOID

Plotting where an accident lies on this simple axis can reveal important information when reviewing accident statistics.

Plotting accidents at work in this way provides a useful additional form of measurement, with clear management implications. Using this axis to plot the lightning strike accident, we can see that the accident was very serious (fatal), but would have been very easy to avoid. Although there is a degree of subjectivity when analyzing an accident using this axis, it should allow a manager to identify and take action regarding incidents that follow a similar pattern of high levels of seriousness (or potential seriousness) in terms of outcome, yet are easily avoidable.

Returning to the lightning strike incident, it's important that we take some workplace-related lessons from what happened, and I'd like to take those two key points I made earlier about our inability to break out of routine and our tendency to ignore clear and obvious warnings.

Many of us like routine, and we operate more successfully with well-defined routines and enjoy having structure to our lives. Getting up at a regular time and eating, working and sleeping at regular times may all form parts of our routine. We also develop routines at work, with regular breaks at certain times, for example, and regular start and finish times, and we develop set routines for tasks that we perform frequently. Routines provide order, and many of us like our lives to be ordered.

There are circumstances, though, when we need to break our normal pattern of routines. In the example mentioned above, stepping outside the daily routine for meal times and walk start times could have resulted in missing the storm altogether. There are other potential benefits to breaking routines. Some people feel more empowered if they can break an established routine; they have more flexibility and life can feel less monotonous.

From a health and safety perspective, being able to break routines and exhibit appropriately flexible behaviour at the right time is one of the key factors in reducing certain types of accidents and incidents at work. When circumstances change, our behaviour and operational procedures need to change in parallel. Sometimes, this may be built into a set of procedures that people have been trained to follow, but in other circumstances, a more free-thinking, problem-solving approach may be required. This isn't something that a workforce can do automatically, though – just as training is required to establish a set of systems or procedures, training is also required to enable people to respond appropriately to a set of changing circumstances, as discussed in the section on free-thinking hazard identification.

Ignoring the obvious was also a contributory factor in the lightning strike accident mentioned earlier, and it's a common contributing factor in a wide range of accident scenarios. Perhaps people just don't want to contradict more experienced or more vocal opinions, even in circumstances in which there is clear danger.

7 Black Box Thinking

Much has been written about black box thinking, sparked by the way in which the aviation industry set about ensuring that the lessons learned from air crashes and near misses were dissected, disseminated and applied so thoroughly. Matthew Syed's book *Black Box Thinking* goes on to argue that success can only come through failure, through admitting mistakes, changing one's perspective and consistently learning from these failures and from the failure of others; after all, you won't live long enough to make every mistake yourself!

It's a simple hypothesis: accept that you make mistakes, learn from them and share that learning with others. The reality is somewhat different, though, because many of us don't like to admit that we got it wrong. We don't like admitting it to ourselves and we certainly don't like admitting it to others – particularly those who we view as our superiors in a hierarchical sense. We're also prone to failing to learn from other people's failures and ignoring their experience-based advice. My son was replacing the cylinder head cover on his car recently, and as the inside workings of the engine appeared, I urged him to put a cloth over the opening so that nothing could accidentally fall down into the engine. He didn't, and I tried really hard to avoid a 'told you so' moment when 20 seconds later he dropped a bolt into it. Luckily, he extracted it before it fell deep into the void, where it could have caused catastrophic failure. I said nothing, a cloth was placed over the opening and hopefully a lesson was learned.

The crucial lesson, though, is not that he should place a cloth over the engine to prevent anything accidentally falling into it, but that he should in the future listen to advice based on the failings of others (me in this case – I've a lot to offer in this respect!).

I referred earlier to a study commissioned by the Department of Health and Social Care revealing that more than 200 million medication errors are made in the NHS every year that could lead to thousands of deaths.

In response, Health Secretary Jeremy Hunt said that 'part of the change needs also to be cultural, moving from a blame culture to a learning culture so doctors and nurses are supported to be open about mistakes rather than cover them up for fear of losing their job'. Pharmacists will be given defences if they make accidental errors, rather than being prosecuted, to ensure the NHS learns from mistakes and 'builds a culture of openness and transparency'.

Dr Andrew Iddles, whose 102-year-old mother was given medication intended for the patient in the next bed, said:

> My mother was given medication intended for another patient – a genuine mistake but it could have had fatal consequences. Whilst we were lucky that the error was recognised and reported, it is encouraging to see the NHS taking positive action to tackle these

potentially devastating errors – above all, transparency is essential so staff can admit to mistakes early without being afraid of losing their job.

This report shocked me. The sheer number of errors is one thing, running at an estimated 237 million per annum, but perhaps the most shocking revelation is how introverted and backward-looking their culture must be, and how far behind some other industries and services the NHS lies in relation to the creation of a positive, no-blame culture that learns collectively from its mistakes.

Learning from accidents and near misses is an absolutely essential part of our approach to reducing accidents and injuries in the workplace, and it's an increasingly valued strategy, but many organizations still do not do this thoroughly and effectively, and there are many reasons for this.

The first and perhaps most important point is that learning through failure has to be seen as a normal and acceptable part of working culture that demands a new type of relationship between workforce and management, and between peers. Management must build a no-blame culture that rewards learning from failure – a distinct shift from a success-rewarding culture that often looks poorly on failure or mistakes, as per the NHS example referred to above.

Integrating real learning from failure also means shared responsibility, ownership and more open communication – all of which are considered positive factors in a more general sense. Management must set a clear agenda from the top down that errors/mistakes/failures need to be reported and learned from. Having said this, although the message has to come over loud and clear from management, devolving responsibility to a workforce is in my view essential, as ownership is much more likely to produce a more positive, communal response.

I developed a health and safety policy for a conservation organization that used the services of hundreds of volunteers to undertake conservation projects. Many of the volunteers were older, retired and very enthusiastic, with wildly varying skill levels and often with no prior experience of working on hazardous tasks. I observed some of the groups at work and admired their positive attitude and willingness to turn out at all times of the year. I also observed many near misses and it quickly became clear that many of the volunteers considered health and safety a nuisance rather than an aid and that many of them wanted to volunteer for conservation work as a way of getting away from the rigorous confines of their past working lives – a sort of freedom from hierarchy, accountability and health and safety. The organization was fortunate, however, in having a very dedicated and skilled team of task leaders, and it was through them that we developed an approach to safety that relied on the volunteers (following training), passing on information about near misses and accidents to other groups. I believe the volunteers responded positively to the idea that they took on responsibility for health and safety collectively rather than have it forced upon them, and they actively wanted to share information that could help their colleagues in other areas.

Conservation volunteers working with RAF volunteers on Whiteleaf Hill, the Chilterns.

One volunteer was involved in an incident in which young ash trees up to approximately 4 metres tall were being cleared from a site on a hillside. Working too closely together, a falling tree missed the volunteer by a few centimetres. The 'feller' gave no warning, and the volunteer who was nearly hit by the tree was working too close to the other. The volunteer spoke to the task leader about it, and that afternoon he compiled a near miss report that came to me. I sent it out to every task leader and volunteer, acting more as a courier than issuing a command. The details were added to the briefing notes that took place prior to every task, and everyone genuinely learned a lesson. I checked on this by visiting other work sites and observed both the addition to the site/work briefing and a practical change in operating methods that ensured volunteers worked further apart to avoid the chance of this happening again.

I made no instruction or order; I allowed the volunteers to be owners of the information and accept the responsibility of disseminating and acting upon it, and I believe that it was the passing on of responsibility that secured such a positive response.

There are a range of additional issues management needs to consider:

- History repeats itself; so does human error and resulting accidents, incidents and near misses. This is a crucial point for managers to be aware of. Finding ways of severing the threads that run historically through safety is essential.
- Targets, data and statistics have their place, but you need to ensure that the desire to achieve certain targets or levels does not prevent your workforce from reporting near misses, incidents and minor accidents.
- Make it easy for people to report when things didn't go as planned and stress the importance of learning from even minor incidents and near misses. My

survey revealed that near misses were not always reported due to the hassle of completing the appropriate paperwork. Make sure a range of reporting methods is available, including verbal, written, mobile/computer-generated forms and so on. Different age groups and people with different learning styles will respond in different ways to reporting methodology. Having just one means of reporting an incident will limit the response.

- Provide training in reporting – this helps to demonstrate how seriously you take it.
- Managers need to understand that a workforce may have a number of reasons for not reporting and for consternation about the process of admitting mistakes. These reasons include a fear of blame or reprisal, not wanting to appear incompetent to peers or management and failing to attach sufficient importance to the process. Only through understanding these fundamental reasons can gains be made.
- Aside from instances of deliberate error or gross negligence, a clear no-blame culture must be established. Management and supervisory staff should also be prepared to admit to and share their failures.
- Understand the difference between different sorts of failure – for example, preventable, complexity-related and intelligent.

8 Risk Perception

After over 40 years working in a variety of hazardous occupations and being involved in mountain rescue, the single point that jumps out at me is that many accidents happen when the level of perceived risk is low. The way we perceive risk is crucial to our efforts to reduce accidents, injury and ill health, and of course it's inextricably linked to free-thinking hazard identification (FTHI).

So what is risk? It has variously been defined as:

- The likelihood that an individual will experience the effects of danger
- A situation or event where something of human value, including humans themselves, is at stake and where the outcome is uncertain
- A situation involving exposure to danger

The concept of risk applies not just to humans, but also to other forms of life, possessions or property, or the environment.

Just to be clear, by comparison, a hazard is any source of potential danger, damage, harm or adverse health effects to something or someone.

Risk perceptions, in turn, are subjective judgments that people make concerning potential harm or loss based on the perceived characteristics and severity of a risk.

Nonfatal injuries to employees by most common accident kinds

(Nonfatal injuries reported under RIDDOR 2016/17)

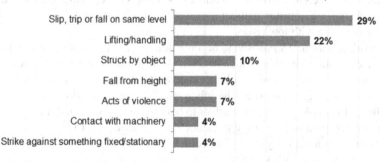

Over 50% of nonfatal injuries reported under Riddor involved slips, trips or falls on the same level or lifting/handling, situations in which the level of perceived risk is generally low.

Both as individuals and collectively, the way that we perceive risk varies dramatically, and a wide range of factors have been shown to impact our ability to perceive risk appropriately, proportionately and accurately. Our perception of risk can be affected by many things, some of which are very relevant to the workplace and some of which are not. Here are some common general principles that are likely to govern risk perception in the workplace.

ANTICIPATED OUTCOMES AND PAST EXPERIENCE

What is going to happen next, and how optimistic are you that the outcome will be positive? Our past experience has a direct effect on anticipated outcomes.

No one would drive without a seat belt if they thought that there was an imminent risk of smashing their head on the windscreen in a collision; however, you might perceive that just driving a short distance to see your mate is actually safe because you've done it many times before without incident, so wearing a seat belt doesn't matter. You don't anticipate a negative outcome; therefore, you can justify your decision based on past experience. You'd be likely to wear a visor when operating a strimmer if you thought there was a high risk of losing your sight as a result of debris striking you; however, you've used a strimmer before lots of times and you can still see, so you're optimistic about the outcome and don't use a visor.

The anticipated outcomes are a result of past experience, but our judgment is flawed through cognitive bias – the making of judgments and decisions based on flawed personal input.

In a workplace or society in which no one prevents the types of behaviour mentioned, it tends to become embedded and normal.

If you have never had to hold a falling climber, and you're sat at the top of a climb belaying your climbing partner as they ascend, the anticipated outcome based on past experience is that they will reach the top safely and you won't have to hold them. For this reason, you may not have adjusted your anchors with enough care and left them slack. If your climbing partner does fall, the outcome is entirely different, and I've witnessed several examples of people being dragged over the edge of a cliff for exactly this reason. The more times you don't have to hold your climbing partner, the more times you operate a strimmer without damaging your eyes and the more times you do that short car journey without a seat belt, the more likely you are to perceive a positive outcome with low risk.

In a work situation, it's possible to become desensitised to risk due to this type of repetitive exposure to it without incident. We've all heard someone say, 'I've always done it like this'. Doing a task and being successful at it repeatedly crystallises our view that it's the right way to do the job, even if we've been taking serious risks. Regardless of the risk, we teach ourselves to believe that it's safe, and it becomes the normal and accepted way of doing the job. Past experience lies to us.

During my main climbing years, I solo climbed extensively. I enjoyed it immensely, although I look back with some horror at this period in my climbing career. The key factors that allowed me to solo climb so extensively were confidence, ability, experience and desensitisation to risk. I started off solo climbing easy routes that were well within my capabilities and climbing longer boulder problems from which a fall could have had serious consequences. Through the 1980s, I started to solo climb more and more, and the gap between my ability limit and the difficulty of the climbs reduced considerably. I'd regularly solo climb ten to fifteen 200-plus-foot rock climbs in the evening after work and I started to solo climbs in the Alps, easy routes on big mountains such as Mont Blanc, the Eiger and Matterhorn and much harder climbs in the Dolomites, some of them 500 m high. Looking back, I was supremely confident and very fit, with a lot of experience to fall back on, but the factor that I am certain influenced my solo climbing most was desensitization to the risks involved. Repeat a dangerous stunt 1,000 times and if nothing ever goes wrong, it ceases to be perceived as dangerous – it's just routine.

On the edge. Perhaps only risk desensitisation allows people to participate in the highest-risk sports.

I stopped solo climbing with a few very easy exceptions in the 1990s, and I now feel genuinely scared when I look at the routes I used to solo climb as a matter of course. So many things could have gone wrong. A loose hold coming away, a small hold fracturing, stonefall, running out of strength, a slippery foothold, a moment of bad judgment. None of those things ever happened to me during this period. The bizarre thing is that I knew that they could happen. I was involved in many mountain rescue call-outs to people who had fallen from climbs after stonefall, when a loose hold had come away or when they simply hadn't been good enough for the climb. I'd seen the fatalities and the life-changing injuries, yet I didn't respond to any of it by stopping solo climbing.

Knowing and understanding this retrospectively has changed my perception of risk generally, and particularly in terms of climbing. Perhaps age and responsibilities had something to do with it, too!

To counter the effects of outcomes being associated with incorrect judgments due to past experience and subsequent misjudgment of risk levels or desensitisation;

- Conduct regular peer-to-peer and external reviews of safety. These don't have to be major safety audits; the idea is to rebalance risk perceptions through a fresh pair of eyes.
- Train workforce and challenge them to identify instances where positive past experiences can lead to inaccurately low risk perception.
- Supervisors and managers should question workforce regarding current work-associated risks and provide a balance cross-check to ensure accurate perception.
- Train workforce in FTHI – risk perception forms part of FTHI training.

Confidence

Confidence that your skills, knowledge and training are sufficient to do the job without negative consequences will increase the likelihood of you judging the task as safe. Confidence is a positive and admired quality, but overconfidence is not, and there's a fine line between the two. Overconfidence can lead to misjudging risk, taking shortcuts and having a blasé attitude and is perhaps most commonly found in more experienced staff.

To counter overconfidence, it's important to:

- Train your workforce in FTHI.
- Ensure that overconfidence is identified and noted as a potential hazard.
- Use experienced staff to train inexperienced staff in hazard identification; this helps re-enforce a correct perception of hazards and associated risks for the experienced workers and provides useful, experience-based training for inexperienced staff.
- Ensure that there is a working culture that allows workforce to question practices such as taking shortcuts on a peer-to-peer basis.

TRAINING, PROCEDURES AND SYSTEMS, PROTECTIVE EQUIPMENT

The core of much health and safety work lies in effective and thorough training coupled to a set of complementary procedures and systems such as method statements and risk assessments. When implemented well, this approach is highly successful, but can it lead to a change in risk perception? Of course it can, because although the mitigation of risk through these factors is considerable, it's not exhaustive. Research shows that using personal protective equipment reduces one's perception of risk. There's a danger that we rely on training, procedures, systems and protective equipment to such an extent that certain risk elements, and particularly risk elements that are connected to change, are perceived incorrectly, with the end result that some risks are underestimated. Counter this by:

- Including this factor in training where and when appropriate.
- Training workforce in FTHI.

TRUST

When we place our trust in the hands of the set of procedures we use at work and the people who have devised them, our fear is diminished and our risk perception is lowered. The more we trust the people informing us about a hazard or presenting us with a solution to a hazard, the less afraid we are. The less we trust the people informing us or the people or process determining our exposure to a risk, the greater our perception of risk. I'd trusted my climbing partner implicitly to set up the belay system to lower me down a 40-m quarry face, yet due to built-in FTHI processes, I still checked, and found my trust had been misplaced. That experience simply increased my levels of vigilance and made me realise how important FTHI is in supplementing even the strongest levels of trust.

Mutual trust, that is, between workforce and management/company is highly beneficial, allowing a better flow of information each way and increasing levels of safety.

- Develop trust between management/company and workforce and encourage a two-way flow of information relating to health and safety.
- Train staff in FTHI.
- Through training, stress the importance of taking individual responsibility for all aspects of health and safety in the workplace.
- Make workforce aware of the link between increased trust and diminishing perception of risk.

JUDGMENT OF THE FEAR FACTOR

The worse the potential outcome from a risk, the more afraid of it we are; the higher the perceived risk, the more cautious we will be. Which is more dangerous, skydiving or swimming? I'm guessing that skydiving is most people's choice, yet you're twice as likely to die swimming compared to skydiving, according to statistics from Germany. In the construction industry, this is to some extent backed up by the statistics, which demonstrate that almost half of the nonfatal injuries were caused by slips, trips and falls at the same level or through lifting, handling or carrying – none of which has a significant fear factor attached to it and therefore carries a perceived low level of risk.

- Provide training to illustrate that low levels of risk perception should not result from low levels of fear or concern in tasks.

Acute or Chronic?

We tend to perceive risk differently in situations that can have a sudden and catastrophic effect, such as an aeroplane crash, in comparison to slow-acting, chronic risks such as heart disease. In reality, the chances of dying in a plane crash are 1 in 11 million, but the chances of dying from heart disease are approximately 1 in 4. Our perceptions of the risks involved in flying in comparison to our perception of the risks created by diet, smoking or lack of exercise do not reflect that ratio.

In the workplace, we need to ensure that employees understand this effect and that equal and proportionate consideration needs to be given to the sorts of slow-acting risks that may impact the longer rather than shorter term. An example could be repetitive bending to lift cement when bricklaying that could lead in the long term to back problems. The risk is perceived as low, as there is no sudden or catastrophic effect, so few measures are adopted to combat the risk until it's too late and the back problems start.

- Management should be aware of the need to link appropriate risk perception to slow-acting, long-term risk.
- There's a clear link with well-being, so integrate this type of risk perception failure with improvements in well-being.
- Train staff in FTHI.

SETTING AN EXAMPLE

Employees look to supervisors and management to set an example. Their commitment towards safety is reflected in the various policies, procedures, training and systems that they implement. These visible forms of support from management may affect employee perceptions of risk. Placing workplace health and safety at the top of management priorities and establishing their position there presents a strong message to the workforce that will be reflected in the majority of their individual behaviour. If management places less importance on health and safety, the workforce is likely to respond in kind.

- Management must set an example and place health and safety at the top of their priorities.
- This example must be re-enforced and embedded over a sustained period.

AWARENESS/NEWNESS AND CHANGE

When a particular risk is given a high profile, either within or without the workplace, our perception of that risk is greater. When the news covers terrorist attacks, particularly if there is more than one in close time proximity, people start to wonder if it's safe to go into city centres, and their perception of risk is magnified by the high profile and newness of the incidents. Awareness doesn't just come from the news media. As individuals, if we've recently experienced something bad, such as the death of a friend through heart disease, or we've been involved in a car accident, awareness of that risk is greater, and so is our fear.

Change and newness increase risk perception. When Sweden switched from driving on the left to driving on the right in 1967, the change precipitated a marked reduction in the traffic fatality rate over an 18-month period, after which the rate returned to its previous values. Drivers were more cautious, as they perceived the risks to be greater, changed and newer. A similar trend was observed when Iceland changed from driving on the left to driving on the right, affirming that trend. Similarly, accident rates gradually returned to previous levels.

In a workplace environment, one could foresee a similar situation developing when, for example, a new construction project is undertaken at a new site. Aside from early teething problems, perception of risk is likely to be at its highest early in the project, when awareness of safety is at its highest, whilst over a period of time, we are likely to observe a gradual change to lower risk perception (marginal loss).

Combating the effects of a reduction in risk perception due to these factors relies on:

- Management ensuring long-term priority is given to health and safety
- Regular re-enforcement of health and safety goals
- Regular peer-to-peer and external auditing/support
- Training and awareness of marginal loss
- Training in FTHI

Incident Effect

If an incident occurs in which a person is injured or killed, our perception of the risk involved in the hazard that caused the accident will be heightened for some time, even though the risk remains the same. This effect can be magnified if the person involved in the accident is known. This is the converse of reduced risk perception due to continuous exposure to risk without incident.

From a management perspective, perhaps it's important to make sure that in such cases other risks are not ignored at the expense of worrying about the particular hazard that caused the accident. Additionally, effective use of near miss reports can help keep your workforce 'on its toes' and heighten certain risk perceptions.

Perceived Benefit

If we perceive a substantial benefit from a certain risk, we may downgrade our level of risk perception. For example, if a few shortcuts that carry higher risk allow a job to be completed early so we can finish early and get to the football match, we'll tend to ignore risks that we may otherwise have perceived to be too dangerous. Trading benefit for reduced risk perception is generally not a good idea, though.

I was once involved in the rescue of a gentleman who had tried to take a shortcut back to his car from one of the Langdale fells. Unfortunately, the shortcut he selected started easily enough, but gradually increased in steepness until he ended up perched on a tiny ledge above a 45-metre drop on Blea Crag, above the idyllic Blea Tarn. I climbed up to him and discovered that there was no way to provide an anchor to make the two of us safe, so we just perched on the tiny ledge for 45 minutes while a rope was anchored 50 metres or so above us and eventually lowered down. The ropes weren't long enough and there were no suitable anchors in the vicinity for us to lower the gentleman to the ground, so we decided to haul him up to the top. I climbed alongside as he was unceremoniously winched up the rock face. As a nonclimber and despite my best efforts, he reached a point at which he simply gave up and slumped into his harness. The process was not particularly kind to certain parts of his anatomy, and after eventually walking him the safe way back down to the road and his car, his departing words were 'I'll never take a shortcut again!'

Occasionally shortcuts pay off, but are they worth the risk?

It's possible to manage the perceived benefit versus risk issue through systems and policies – for example, by requiring permission to deviate from standard practice and providing awareness training in the dangers of taking risks due to perceived benefit.

Many other factors can influence our perception of risk to some extent, so the list above is not exhaustive. For managers, it's very important to make sure that you have these discussions with your staff, so that they understand the principles of risk perception and how they affect their resulting actions. Training and communication can help a workforce make and maintain accurate appraisals of risk. Bearing in mind that risk perception issues lead directly to a significant percentage of accidents at work, this is a conversation that's well worth having.

Finally, here is a summary of my personal experience of risk perception:

- In outdoor activities, many more accidents occur when the perception of risk is low. When you perceive risk levels to be high, you are more cautious and take appropriate measures. For all the factors mentioned previously and many more besides, one of the keys to safe working is never to consider perceived risk as low. Delete the phrase from your vocabulary.
- The period immediately after you move from a situation of high perceived risk to a situation of low perceived risk is most dangerous.
- Don't allow your workforce to get sucked into the trap of risk desensitisation.
- Remember that factors such as culture, gender, ethnicity, education and socioeconomic background may have significant impacts on risk perception.
- Individuals approach risk with different attitudes and backgrounds, opinions and experiences. We don't all approach and perceive risk in the same way, so our training, education and policies must take this into account.
- A risk perception analysis can be a very worthwhile exercise that will help to level out the differences that a workforce is likely to exhibit.
- FTHI helps to ensure that risk perceptions are accurate and proportionate.

The best established results of risk research show that individuals have a strong but unjustified sense of subjective immunity.

Douglas, 1985

9 The Problem with Experience

If you needed a job done on your house, or you needed a childminder, a mechanic, a fitness instructor or a gardener, money was no object and I offered you a choice between an experienced person who had done that job/performed that role for many years and an inexperienced person who'd just come out of college, I'm willing to bet that most people, most of the time, would go for experience. We perceive that we will get a better job done and probably in a shorter time. We also believe that experience will result in fewer things going wrong. We also naturally extend this principle to safety matters and we associate more experience with higher levels of operational safety.

Let's be clear about what we mean by experience in the context of safety at work. What we're referring to is the process and result of acquiring knowledge and skills from performing a task repeatedly and learning about it intellectually and physically. There's an element of subjectivity in this, too – who is the most experienced driver, the one who has driven up and down a quiet stretch of the M6 each day for 10 years, or the one who has driven half the miles, but on busy A roads and country roads? They'd both say that they were experienced drivers.

In the late 1980s, I went to Colorado one winter to ski with my best mate, who had landed a job for the season at Copper Mountain Ski Resort. On the first day I awoke to −20°C and vivid blue skies and couldn't wait to get started. My friend was working, so I elected to try and hitch a ride through to Loveland Pass, from where I would head out on an off-piste ski tour. I managed to get a lift down the main highway, and stood for a while, shivering at the foot of the pass, before a battered old pickup stopped and offered me a lift. The driver motioned me to put my skis in the back, which I noted was already full of a huge assortment of ski gear. I jumped in the front and we set off up the pass. The driver was Phil Huff, a local skier, and, keen for some local knowledge, I told him what I was wanting to do that day, but Phil wasn't too interested in that – he had recently invented a new sport and he wanted to tell me all about it.

Phil told me that, along with a friend, he had invented a special sail that allowed skiers to use wind power to ski uphill – Upski. As I took in his description of how his invention allowed skiers to use the wind to speed uphill, like having your own personal ski tow, I experienced one of those so-called light bulb moments, when your brain suddenly powers up, and you realize that something profound is happening. Phil met a few friends at the summit of the pass, including Upski legend John Harrington, and introduced me, before giving me a quick lesson in how to use the Upski. As I tried to take in everything he was saying, my eyes had already been drawn to one side, where the others had unpacked and inflated their Upski canopies, and were already snaking at speed up the hillside, their multicoloured Upskis shimmering in the bright sunlight. It looked, quite simply, awesome.

Overconfidence and the result of high levels of experience in similar skills caused two potentially serious accidents relating to using Upskis.

I was soon hooked up to my own Upski, and as I watched the wind fill the voluminous nylon canopy, I started to feel the pull of the wind and proceeded to fall flat on my face. The second time, I knew what to expect and though neither as graceful nor competent as the others, I whooped with delight as I felt the wind power me across the snow. I quickly learned to lean back in the harness and use the ski edges like a keel to take a line to the side of the wind. At the top of the slope, we deflated the canopies and trailed them behind us as we sped back down. Several exhausting hours and many falls later, I was totally hooked. The possibilities seemed endless. The Cumbrian fells and the Scottish mountains had lots of snow and a seemingly endless supply of wind, surely perfect terrain for an Upski. By the end of my holiday, I had become the European agent for Upski and set about promoting the concept back in the United Kingdom. I supplied Arctic and Antarctic expeditions with Upski equipment, it was featured on BBC's 'Tomorrow's World' and I even enjoyed a couple of days in the Cairngorms with Blue Peter and the late Karen Keating.

I was determined to let everyone know just how wonderful the Upski experience was, and that spring I had arranged a meeting with the head of the ski school at Glenshee, a local legend named Gustav Fischnaller. We met late in the afternoon as the pistes were quietening down, and a brisk westerly made for perfect conditions. I showed Gustav how to use the controls, how to activate the emergency releases and how to inflate and deflate the canopy. We had a trial run, then sped up and down a lovely section of off-piste snow. The weather closed in a little, the westerly wind picked up and it wouldn't be long before dark, so Gustav set off on a final run. He disappeared into the mist-shrouded hilltop, where the white of the snow and the grey of the mist merged into a gloomy steel grey. I waited, but there was no sign of Gustav's return. I inflated my canopy and went to investigate, instantly aware of how much more powerful the wind was. I edged towards the mist and followed a set of ski tracks into the gloom. I soon reached the summit plateau, where the wind had scoured the snow, leaving hard ice and exposed granite boulders. That's where I first saw the blood, dappled in the snow behind one of the rough granite lumps. I deflated my own canopy completely, and it still pulled me – it was a fight to stay upright and pack it away. The wind pulled at me as I followed the tracks and blood across the hard snow. I came across some ripped clothing and more blood and I began to fear for Gustav. The ski tracks disappeared, but I simply followed the trail of blood and ripped clothing until the slope changed abruptly and started to descend steeply. A few minutes later, I came across Gustav laying in the snow, barely conscious and clearly in a bad way. I wrapped him in the Upski sail for protection and started back up the hill to get help. My legs felt like jelly and it seemed to take forever to reach the summit and my skis. It was almost dark now, and although I knew that if I simply went directly into the wind it would take me the right way, the whole scenario felt very serious. I reached the slope where we had originally ascended, and as I came out of the mist, I spotted the lights of a piste basher heading up the piste just half a mile away. I skied down to it and told the driver the situation. He alerted the main centre and a rescue party set off from there while we trundled up the hill, the piste basher making light work of the conditions. We eventually reached Gustav again, and shortly after, a group of rescuers appeared out of the darkness and he was carefully strapped to the stretcher and evacuated, first by carrying and then in the piste basher. He was seriously injured and it was a difficult, sobering situation. I visited Gustav in hospital a couple of days later and was delighted to find him conscious and improved, though clearly shaken by the experience.

I examined the Upski equipment immediately after the accident, and reflected on that afternoon at some length to see what lessons could be learned.

During the late afternoon after meeting Gustav, the winds had gradually strengthened, leading to a significant amount of blowing snow, or spindrift as it's called. Upon examination, I discovered that the nylon sheaths that encase the four pulley systems that allowed the Upski canopy to contract and expand had become twisted, knotted and jammed with compressed snow. The snow was only just below freezing temperature and the blowing spindrift had, over time, found its way into the nylon sheaths and compressed into tight balls, which wrapped around each other and jammed. When the normal release was used, the canopy failed to deflate and continued to exert a pulling force on Gustav. He in turn ignored the second and third safety releases that I had showed to him, and reverted to his training and experience with paragliders. He tried to deflate the Upski canopy in the same way as you would depower a paraglider, by pulling the bottom lines towards you. The Upski works on a different principle, and this method failed. A strong gust of wind had exerted huge force on the canopy and Gustav had been helplessly pulled along the boulder-strewn and ice-clad plateau until he was deposited on the steep hillside in the lee of the wind. Had he simply activated either of the two emergency releases, the accident would not have happened.

This accident exemplifies some of the common issues relating to safety that apply to experienced people.

- As Gustav and I were experienced skiers, mountaineers and paragliders, we were overconfident.
- We failed to take into account the gradually strengthening wind speed and deteriorating visibility, and failed to move down the mountain to a better

location, poor decision making that resulted in part from overconfidence and in part from distraction.

- Gustav failed to follow the emergency procedures that I had shown to him, and reverted by habit to his knowledge of a similar, yet very different, sport, paragliding.
- In terms of training Gustav, his exceptional confidence and skill levels in skiing had led to him being able to use the Upski easily and I had overestimated his ability to translate the training into action in the event of an emergency. I had also underestimated how much training in emergency procedures was actually required, even with such an experienced person.

In the early 1990s, the world-famous explorer, Sir Rannulph Feinnes, and his trans-Antarctic colleague Dr. Mike Stroud contacted me. They were interested in using the Upski canopies to help drag their sleds on their Antarctic crossing that started in late 1992. I agreed to loan them some equipment, but insisted that they needed to be trained in how to use it. Due to time constraints on their part, that training consisted of 10 minutes in a lay-by near Kendal with Dr. Stroud and me providing some comprehensive written instructions.

Some time after their return, I was invited to the Royal Geographical Society to hear Feinnes and Stroud report on their incredible 95-day unsupported trek. During the lecture, it was disappointing to hear that they blamed the Upski for dragging them into a crevasse, and gave it a bad report, although I later found that they had also had plenty of good use out of it. Some time later, Feinnes dumped the Upski in his porch in London for me to collect, which I duly did. On examining the Upski the following day at home, I found that it had been altered significantly – changes that led directly to the problems they had encountered. Around the edge of the circular canopy, a web of mesh is interwoven with the lines that attach the canopy to the user. This web serves a vital purpose, to prevent the canopy from inverting – basically blowing itself inside out. If the canopy does blow inside out, it doesn't deflate fully and emergency releases need to be deployed. Feinnes and Stroud had cut off all of this mesh – partly to save weight and partly as the mesh was sometimes getting caught in satstrugi, wind-created pillars of snow that can be anything from a few inches to a few feet in height. In removing the mesh, the canopy would have readily inverted, leading to the situation described in which the explorers were dragged into a crevasse. Additionally, the failure to understand both the function and necessity of the release mechanisms led to incorrect attachment of the Upski and subsequent difficulty in operating either of the two emergency releases.

This story echoes some of the themes from the previous accident. In this case, we had two explorers who were so supremely confident, so convinced of their abilities and invincibility, that they didn't take the training I offered seriously, and were happy with 10 minutes in a lay-by and some written instructions that I doubt were read. Their overconfidence led them to ignore the need for training, in turn leading them to make dangerous alterations to the equipment that could have killed them. Perhaps in hindsight I should not have loaned the equipment knowing that the level of training provided was inadequate, but the pressure of dealing with such well respected, well-known and confident individuals skewed my judgment.

Some years later, I received a letter from Sir Rannulph while he was in South America, waiting to head to Antarctica once more. In the intervening years, kites had been developed that operated on a ram air principle and offered more versatility than the Upski, and Sir Rannulph had seen them in action. He wrote 'Any well prepared boy scout with one of these kites could cross Antarctica these days...'

I was at a garage I know the other day watching my car being put through its annual MOT roadworthiness test. I know the mechanic – a grizzled old guy who's been in the motor trade since he was 16 – and we started chatting about safety as a result of an inspection of the lifts and jacks that was taking place at the same time. There were a couple of young lads listening in to the conversation and I got the feeling that safety didn't interest them too much, at least until the old guy started to tell them about his experience with an angle grinder. He'd been working with the grinder on and off one afternoon, when it suddenly kicked back at him, flew upwards and sliced into his neck and throat. He went on to describe how he'd put a cloth over the wound and walked down to the local chemist and asked for some steri strips. The chemist asked to see what he was going to use them for so she could provide the right size, whereupon he peeled back the cloth on his neck. The chemist fainted and he ended up in hospital, the grinder having missed his main artery and windpipe by just a few millimetres. His thick, silvery beard now hides the scar, but he described how it took many months before he could use a hand-held grinder again, and even now he feels very uneasy using one.

Used in the right way, experience is an invaluable asset, offering us an insight into many of the things that can go wrong at work and in life in general. By contrast, experience is also no guarantor of increased levels of safety, offset as it so often is by factors such as overconfidence and desensitization to risk.

Experienced people have as many accidents as inexperienced people, and we need to manage experience in the workplace as much as inexperience. Statistics from the Castle Climbing Centre show that experienced climbers are equally likely to be involved in accidents, incidents and near misses as inexperienced climbers. Some studies of accidents in the agricultural and construction sectors have shown that young workers are involved in fewer accidents than older, more experienced workers, while other studies have shown little difference. Some research has shown that the level of seriousness of injury and number of fatalities as a result of accidents with young, inexperienced workers is lower than that for older workers – perhaps a reflection of how younger people are more resistant to impact and trauma. After looking at much research and anecdotal evidence, my conclusion is that more experienced workers are equally likely to be involved in accidents, incidents and near misses as inexperienced workers, but that the causes and results may be substantially different, and industry-to-industry differences may be appreciable.

THE PROBLEMS WITH EXPERIENCE

EXPERIENCE OF WHAT?

If someone has performed a task successfully many times, he or she will say he or she is experienced at that task. This usually gives peace of mind to all concerned.

However, the person may have no experience of things going wrong with that task and of what actions to take in that circumstance. Therefore, we must differentiate between experience of things going right and of things going wrong. Training must provide some focus on managing the task when something goes wrong, in addition to training in correct application.

Concept of Experience Providing a Buffer against Accidents

Experience may provide a buffer against certain types of accidents, but popular perception is that experience is a buffer against all ills. It's not. Statistics show that experienced workers are just as likely to have accidents and injury as inexperienced workers. We need to understand why this is and take practical steps to guard against this issue.

Transference of Skills and Knowledge to Broadly Parallel Situations Might Not Work

Experienced people often try to transfer their skills, knowledge and techniques to broadly parallel situations – but this might not work; Gustav Fischnaller reverted to his experience in the vaguely parallel sport of paragliding to try to deflate the Upski canopy that nearly pulled him to his death, ignoring the training he'd just received and habitually trying a technique that was never going to work in this new situation. Acknowledging that skill transference might not work and being prepared to understand new, but broadly similar, tasks is essential. This type of behaviour links with habit and routine, and training needs to be provided so that workers understand the dangers of attempting skill transference and of reverting to habitual actions.

Being Blasé or Complacent, Switching Off or Running on Autopilot

These are classic issues with experienced people. The task has been performed so many times successfully that nothing can go wrong – can it? There can be very few of us who have never experienced issues with complacency, with switching off mentally as we go through the same process in a task yet again. These issues can lead to a lack of concentration and focus, and when you're not 'switched on', you're vulnerable.

Overconfidence

In the examples I discussed earlier, I believe that Gustav, Sir Rannulph, Dr. Stroud and myself were all guilty of overconfidence. Overconfidence is very dangerous; it skews our judgment of risk, blunts our observational skills and dulls our response to change. It's essential to teach workers the dangers of overconfidence and to preach to them the importance of free-thinking hazard identification, constant vigilance and continuous assessment.

RISK PERCEPTION

I've discussed risk perception elsewhere in this book, but one of the classic issues with the way that we perceive risk is that when we have performed a task many times without incident, we perceive the risk as lower than it really is, thus laying ourselves wide open to problems. It's essential that experienced workers be trained to recognize the dangers posed by desensitization to risk as a result of lengthy, uneventful exposure to it.

ASSUMPTION THAT OTHERS HAVE EQUAL KNOWLEDGE

It's common for experienced people to make assumptions about the experience, knowledge and ability levels of others; the importance of not making these assumptions must be stressed, and forms part of the concept of continual assessment.

MANAGING EXPERIENCE

GET RID OF IMMUNITY TO RISK

Workers must understand that experience does not equal immunity to risk, but perhaps means the opposite. Training should focus on risk perception and analysis, free-thinking hazard identification and working on strategies to raise awareness of risk among experienced workers.

LEARN THAT EXPERIENCE ISN'T ALWAYS RIGHT

Experienced people are often right, but don't have a monopoly on it! An inexperienced person might spot something that an experienced person has not seen, yet inexperienced people often don't communicate that, as they feel the experienced person must be correct. Management must create and establish an open culture that values the contribution of younger and inexperienced workers, and in which acting in everyone's interest, even when wrong, is valued and accepted. An inexperienced worker perceiving something is being done dangerously must feel free to ask an experienced worker why a task is being performed in a certain way. The experienced worker might learn from the observation, or the inexperienced worker may learn that the task is being done in that way for a reason he or she was not aware of and that the risk is being fully managed. Whichever way round, someone benefits, but management has to work hard to create and sustain that culture.

USE EXPERIENCE TO WARN OTHERS

Use experience positively to pass on knowledge about potential safety issues to others. Experienced workers can usefully relearn skills – that idea of going back to basics. One good idea is to use experienced workers in an instructional capacity, which serves more than one purpose. Although the impetuous nature of youth and its skewed desire to learn lessons its own way rather than listening to the voices of experience is a problem, there's also no doubt that in the right context and situation, learning from the experience of others is a critical part of managing health and safety.

GUARD AGAINST COMPLACENCY AND OVERCONFIDENCE

Stress the importance of guarding against complacency in experienced workers. Encourage other less experienced workers to question actions and methods, and encourage experienced workers to accept questioning. Train workers in free-thinking hazard identification, continuous assessment and constant vigilance.

Experience can be dangerous!

10 A Personal Perspective
Part 2

I worked with the Lake District National Park service initially as a labourer mending walls and repairing paths, then as a Volunteers Organiser and later as a National Park Ranger, managing the busiest section of Britain's premier National Park. During this period (1979–1989), I recall very little attention being paid to health and safety in relation to our working environment. We regularly went out in harsh winter conditions on our own onto the high mountains and undertook a significant amount of what we now refer to as lone working. Our radios may or may not have worked, mobiles were still the size of suitcases and we frequently didn't have any official way of letting people know where we were or what we were doing. We carried people in the back of our vans, where they bounced around with the tools and equipment. We humped heavy fence posts and gates over marshes and up mountainsides and heaved giant boulders to build paths. During the whole of this period, I never even heard of the phrase 'risk assessment', let alone completed one.

Mountain and water rescue were important components of our job, and I recall solo climbing difficult rock climbs to assist 'cragfast' scramblers or climbers on rescue operations. One summer afternoon, a 16-year-old boy had somehow got himself stuck on a small, sloping grass ledge on a 500-foot-high cliff called Pavey Ark, in Langdale. Two hundred feet from the ground, the lad was terrified and worried he was going to fall. I decided that rather than wait until we had abseils rigged up at the top of the cliff, I would climb up to him, such was the precarious nature of his situation. After easy scrambling, I had to climb the first pitch of a Very Severe (VS) rock climb, not made any easier by the fact that I had no rock climbing shoes with me and had to borrow some walking boots to climb in. To get from the climb to the young lad involved climbing across damp, sloping ledges; heathery slabs and moss-encrusted cracks. It was a nightmare, but I was so focused on reaching the boy that all other thoughts were banished from my mind. He was stood on a damp, sloping grass ledge, from which a slip would have resulted in a 150-foot vertical fall, and he was terrified. I reached him safely, and as I stood with him on the ledge, I have to say I didn't feel too happy myself! There was no way of protecting us both until a rope came down from above, so for half an hour, I kept him as calm as I could in the circumstances, until at last, a rope snaked its way down to me. Once attached, unknown rescuers way above us held the rope and I leaned back over the void with the lad tied securely to the rope next to me. In other circumstances, the 150-foot abseil might have been quite enjoyable, but not when you're trying to control a highly emotional teenager. I'm not sure who was more relieved to reach the ground, but once safely down and following a short, rocky descent to the security of the path at Stickle Tarn, we all relaxed, and discovered the reason why the youth had so nearly lost his life. Quite unbelievably, his mate – part of a small youth group walking up the mountain with an inexperienced and unqualified leader – had bet him £1 he couldn't climb to the top of the cliff!

Almost everything we did would now require a completely different set of working procedures, and our culture was in some ways akin to that of the farmers – we just got on and did what needed doing. I recently spoke about the lack of any sort of health and safety culture during that period to one of the supervising Area Rangers from that time, Pete Rodgers. He agreed with me and said, 'It just hadn't filtered down. We took on board basic safety equipment and techniques for operating chainsaws, but nothing else'. The National Park Authority was effectively a government organisation, and our lack of knowledge about health and safety matters perhaps illustrates how long it could take for policies and legislation to work their way through to a workforce. Nowadays, advancements in communication and an audience that is already in possession of a strong health and safety culture make for a much faster and more effective rolling out of new regulations, ideas and policies.

The interesting fact, though, is that throughout this period, despite frequently working on difficult and often dangerous tasks, frequently lone-working and with not a single health and safety policy or risk assessment in sight, neither myself nor Pete recall any accidents that involved the National Park Ranger service. We were acutely aware that we were often performing hazardous tasks and frequently operating in dangerous environmental conditions. I strongly believe that the fact that we were so acutely aware of this enabled us to work in a safe manner without the benefit of a set of procedures and systems and without performing risk assessments. This reinforces my belief that when we are knowingly working in hazardous environments, our perception of risk and our reaction to it are far more highly tuned than they might be in other working environments.

11 Free-Thinking Hazard Identification

So, what does this mean? The hazard identification bit is easy, it's just as it says. The free-thinking bit is more difficult, though, as it's a conscious, progressive and very thorough mental process that in most people requires switching on, rather than being inherent or automatic.

If I take a walk from my house up to the road and across to the fields on the other side, I'd probably have a quick look to see if any cars were coming, and that would be about that in terms of concerning myself with hazards. However, if I switch into my free-thinking hazard identification mode, I start to see many other potential dangers in addition to the obvious one of the cars on the road. I notice the small step down from my kitchen and the low headroom. I look at the hallway rug and check that it's not upturned to present a trip hazard, and I make a quick visual check to see that the boys haven't left shoes/car parts/dead mice or anything else on the floor to slip on or trip over. I open the door and note, then step carefully over, the lip at the doorstep. Before walking up the steps, I check there's nothing on them and that they are dry. If they are wet, I make a mental note that they will be slippery and place my feet carefully. At the road, I stop and check not only for cars, but also for cycles, which come past at speed regularly and make no sound. I cross the road with care, and when I reach the gate on the other side, I check there is nothing to trip over as I pass through it and open the latch carefully to make sure I don't trap my fingers on it. Yes, it's sad!

Putting yourself into FTHI mode is in effect magnifying and sharpening your ability to identify static hazards, and raising your awareness and response to change. In FTHI mode, you don't look on the surface and identify the obvious hazards – you peel away the layers and really understand the depth of what can go wrong. It's a state of mind that can be learned, practiced and honed. It demands concentration, focus and a strong will, and it's helped by being alert and having positive well-being.

Only a small percentage of climbers solo climb – I was one of them. I solo climbed 500-metre vertical climbs in the Alps, thousands of rock climbs in the United Kingdom on the mountains, on gritstone outcrops and on imposing sea cliffs. Looking back at it, I shudder at how fine the margins sometimes were between sticking to the rock and falling off it – between life and death. Madness! Yet it was also extremely carefully calculated and executed. Everything had to be right – the weather, the temperature, the clothing I wore and most importantly the feeling of confidence and intensity of concentration. The risk levels were open, unambiguous and unimaginably high, but once I was climbing, the intensity of concentration and depth of focus was unlike anything I have ever experienced. I likened it to focusing the whole of your physical energy and mental capacity onto the end of a pin.

This state of heightened awareness is not unlike the feeling of concentration and focus we're aiming for in our state of free-thinking hazard identification.

GETTING INTO THE FREE-THINKING HAZARD IDENTIFICATION ZONE

When a leading football team prepares for a game, as individuals, they have to switch from their normal mode to game mode. Normal mode is relaxed, chatting to teammates, perhaps texting a friend or thinking about the latest Ferrari that's being delivered to them the next day. However, they have to find a way of separating that mode from the intense focus and concentration that's required of them once the game starts. In order to do this, they have to use a psychological trigger that throws the switch to 'on' when you 'cross the line'. Crossing the line could be exiting the changing room, transferring from the tunnel onto the pitch – the important point is that there is a defined point at which things change.

We have to do this if we want to adopt an FTHI approach. We need to cross an invisible line beyond which we enter FTHI mode. This line can be set wherever you want it to be, but it must trigger the response of switching from normal mode to FTHI mode, so it needs to be a trigger that you will be exposed to every working day. You'll all have seen workers drift into work, still chatting about yesterday's football or netball, preparing a few tools and so on, clearly not concentrating fully on work. There's no harm in chatting about sport, but that needs to become a secondary issue to the primary focus of being safe. Importantly, this trigger response, the change to FTHI mode, should also take effect after each break.

Before setting off on a climb as a lead climber, I'd normally chat light-heartedly with my climbing partner and have some fun as a way of relaxing prior to the climb. However, there always comes a point at which the switch is thrown, and all attention and concentration become focused on the climb. I used the tying and checking of the knot linking rope to harness as my trigger point. Once that was checked and completed, my focus was purely on the climb.

One of the best ways to kick-start your own FTHI process is to select a task or activity that is familiar, then identify the key hazards associated with it, rather like I mentioned in the example above. Once you've done this, go on to examine every part of the task/activity process in much more depth and see how many more hazards you can identify. Delve deeper and deeper into the process and you'll find that you can identify more hazards than you thought.

Having done this, you should then select a task or activity that is unfamiliar and that you perceive has a number of hazards associated with it. On courses, I use activities such as rock climbing, abseiling, gorge scrambling, paragliding, skiing and so on. You don't have to know the activity, you don't have to have participated in it or have an interest in it. It's simply a good set of wheels to learn from.

For my example here, I'm going to use a very simple activity – top-rope climbing on an indoor climbing wall – and I'm going to focus on just one part of the process – that of the belayer, the person who is operating the safety system. You could watch a video of someone belaying a climber, or you could watch people climbing at a local climbing wall or even look at a series of photos in a book. Whichever way you do it, your first objective is to understand the task that the belayer is performing. Write down a definition, for example:

The belayer controls a rope that runs from the climber, up through an anchor point at the top of the wall and back down to the belayer who has the rope threaded through a belay device. The belayer can use this device to take in the slack rope as the climber ascends and lock the rope off should the climber slip on the way up. The belayer also lowers the climber back down when he or she reaches the top or gets as high as he or she wants to go.

Once you have understood the purpose and nature of the task, you need to go through it and identify the obvious, key hazards. For example:

- The belayer fails to use the equipment properly, the climber falls and the belayer cannot hold him or her.
- The belayer doesn't concentrate and lets the rope slip out of his or her hands after the climber falls.
- The belayer lets go of the rope while the climber is being lowered down after reaching the top, resulting in a fall.
- The climber falls and lands on top of the belayer.

It's normal that your first run through hazard identification in an unfamiliar activity will focus on the more obvious and major hazards and will be influenced by any knowledge you have of the activity, whether direct or perhaps of a parallel nature. It's most likely that you'll also initially question systems for checks such as whether the rope is suitable for climbing and when was it last checked for damage.

The next stage is to break down the main hazard components into deeper levels and start asking the question 'What if?'

Let's take the first hazard identified above as an example – 'the belayer fails to use the equipment correctly'.

Take a look at the photo – you'll see that there are a number of components, including the belayer (human component), a harness, a belay device, a carabiner and the rope. Most people quickly come up with the idea that the equipment needs to be of a certain standard, which of course it does. They also come up with the idea that it needs to maintained and checked regularly, which of course it needs to be. So we move on – the equipment is fit for purpose and is maintained and inspected on a regular basis. The next thing most people recognise is the necessity for the belayer to be trained correctly in the use of the belay device and other equipment components. So far, so good. We then start breaking things down further. Let's look at just two of the items of equipment – the harness and carabiner.

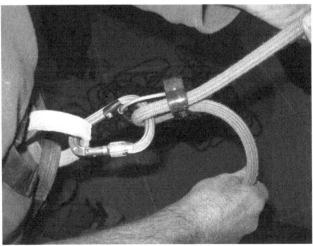

Top-rope climbing at an indoor climbing wall, belay plate, carabiner, harness and rope in use.

The Harness
Is it appropriate for the task?
Does the harness exhibit any wear and tear and, if so, is it acceptable? How do we know?
Who decides?
Where does it tend to wear?
Is there likely to be any wear that is not visible?
Is the harness put on the right way round?
Are the harness buckles fastened correctly?
Can they come undone in use?
How tight do they need to be?
Do I tighten the waist or leg buckles first?
What do we do with the slack belt left over after fastening?
How much spare belt should be left over? Is there a limit?
Is the harness tight enough on the belayer? How do we know?
What if it's not tight enough: what happens?
Does it fit correctly?
Are there different harnesses for men and women?
What position should the waist belt be in?
What if it's too low; how does this affect the centre of gravity?

The Carabiner:

Is the carabiner fastened to the right place on the harness, and what happens if it isn't?
Does the carabiner need aligning in a certain way because it's asymmetrical?
What happens if it's aligned incorrectly?
Is the locking gate fastened or unfastened?
What if it's not locked?
Do the colours mean anything?
How do we know if it's worn to an extent that it might fail?
And so on...

I'm questioning as much as I can and looking for answers to ensure that the physical use of the equipment is appropriate and correct. This is one key part of the FTHI process, and it's very much the product of the relationship between the worker and the equipment.

Beyond this, it's a question of then visualising how the equipment is going to be used and, more importantly, visualising what could go wrong with, for example, changing environmental or working conditions. This leads us nicely to the concept of visualisation and its application as a tool for FTHI.

VISUALISATION

You'll often see climbers at the foot of a route or boulder problem, standing back a little and moving their hands about while gazing intently at the route as if in a trance.

You'd be forgiven for thinking this was some sort of bizarre ritual, but working out the route in advance is actually really useful, and the more experience you have, the more accurately you can predict the moves through visualisation.

True visualisation converts visual information into sensory feeling. While reading the moves on a rock climb produces a series of information, visualisation takes this information and applies it in a much more powerful way.

Visualisation techniques are well documented and are used by most professional athletes. In rock climbing, techniques may involve imagining watching yourself doing the route through to a stage where you're feeling the texture and shape of the holds and each subtle body movement in an extremely vivid, yet imaginary way. Some athletes describe the visualisation process as so intense that when performing a task, they feel like they have done it before. Leading climber Dave Birkett explains that, 'Before I do a new climb, I go through it all time and time and time again. Once I've worked on a route and visualised it so many times, I know I can climb it. You get to a point in your mind where the route's already done. The game's won when you know you can do it even before the ascent. When I get to the route I just see how it feels and I know in the first couple of moves if it's going to go. When Jerry Moffat won the Leeds competition years ago, he used to visualise everything from walking onto the stage, putting his harness on and tying in, waving to the crowd – absolutely everything'.

Dave's climbs often combine extreme difficulty with high levels of danger, so this sort of visualisation process is an essential part of his preparation.

To aid performance in high-level sport, a period of visualisation precedes the action. Here, a climber spends several minutes visualising the moves on the climb before attempting it.

Visualisation techniques for FTHI are important because they allow us to see accidents happening before they have happened.

Peter Shipley, Programme Delivery Director at Tideway, explains an example from the construction industry:

> I heard of a worker using an angle grinder on site to cut reinforcing bars, in accordance with the method statement. He had to stretch into a cage to reach a bar to the point where he lost control of the angle grinder and ended up with a nasty cut on his arm. It could have been much more serious. The focus on the method statement as being the main tool for ensuring safety reduces the focus on FTHI. The operative just didn't imagine or visualise the outcome that happened – he was too busy getting the job done and the box-ticking exercise driven by overadministrative method statements.
>
> A better approach would be to use more visual and much simpler method statements, but to include the FTHI trigger moment at the start of the operation. Perhaps the morning site briefings that all staff have at the start of a shift could include those FTHI trigger moments along with discussion about a range of potential accidents that could happen that day and a reminder for all to stay safe. This can be linked to the concept of perfect day.
>
> Perfect day is a concept our contractors are using that focuses on a one day at a time – let's have nothing go wrong. This idea gets people thinking every morning about

what might happen and what they need to do to have another perfect day. We have charts up at each site with the days on the chart and a smiley face gets put in for every day that passes without an accident.

YOU COULD SEE THAT ONE COMING

There has recently been an advert on TV in which a lady walks through a series of crowded streets and adeptly moves pushchairs, skateboarders, pedestrians and others out of harm's way just before some major accident would have befallen them. It was something of a relief to see the baby in the pushchair not getting buried under a ton of cement, and heart-warming that the skateboarder falls comfortably onto a padded mattress rather than a bone-breaking stone floor. In addition to this mildly entertaining bit of cinematography, the advert exemplifies magnificently the vital concept of seeing it before it happens.

How many times have you heard people say, 'You could see that one coming!' Sometimes it's blindingly obvious that something is going to go wrong and everyone can see it. More often than not, though, it's not so obvious, and no one sees it coming. The classic sideswipe.

In my many years working as a mountain guide and instructing outdoor pursuits, I developed an acute and absolutely vital skill – that of visualising things going wrong before they did, and taking action to make whatever corrections were required. Here are a couple of examples:

When beginners enjoy their first rock climbs, they are usually introduced to it using a top-rope system. A strong anchor is set up at the top of the crag and the rope runs from the climber up to the top of the crag and through a carabiner attached to the anchor. It then comes back down to the ground to the belayer – the person who is in control of the rope while the climber ascends. It forms a sort of pulley system. The climber ties one end of the rope to the harness and the belayer threads the other end of the rope through a belay device, which is under his or her control. This device enables the belayer to take in slack rope from the system as the climber ascends, but also to lock the rope if the climber were to slip, thus arresting the fall so that the climber simply slumps onto the rope. All relatively simple, and you can see this setup at climbing walls and beginners climbing crags outside all over the country.

I'd watched a couple of my instructors set up top ropes as described on a steep little rock face called Lower Scout Crag in the Lake District's most famous climbing valley – Great Langdale. I'd seen the anchors used, their positioning, the fastening of the harnesses and so on – everything was being done perfectly. However, one of the key components of the fall arrest system – the rope – has a number of properties that need to be managed correctly. One of these is that a standard climbing rope stretches to absorb the energy created in a lead climber fall. This is an essential property that reduces the shock loading on all the other components of the fall arrest system in a major fall, but in top-rope climbing, the stretch can become a liability rather than an advantage.

So let's consider what would happen if a heavy adult climber manages to get 6 or 8 feet off the ground and then suddenly falls. Even if the belayer locks the rope off correctly, the rope stretch coupled with the positional change to the belayer due to the pull generated could be sufficient to land the climber painfully back on the ground. The solution is easy – before the climber starts, get rid of the rope stretch by pulling hard on it. Then, if the climber falls, he or she just slumps on the rope, and the risk of hitting the ground and potentially being injured through rope stretch is substantially reduced.

A quick shout of 'don't forget to take the stretch out' was enough to rectify the situation. To put this in perspective, it would have been unlikely that anyone would have come to any harm had I not spotted the error, but unlikely doesn't mean impossible and I'd rather not take the chance.

On another occasion, I was climbing with a friend in one of the large, disused quarries in the Coniston area of the Lake District. We'd done a couple of climbs and things had gone really well; it was a beautiful evening, and we decided to do one final climb by top rope. The easiest way to the bottom of the quarry was to abseil down, but as I was coming straight back up again, we decided that we'd set up a belay and lower each other to the bottom, the difference being that when you abseil, the control is with the abseiler, but when you're being lowered, the control is all with the other person. We chatted away and I stood at the edge of the quarry, ready to be lowered the 40 metres to the bottom, when the FTHI visualisation process kicked in automatically. I looked down at the descent and noted any issues such as sharp rock edges or loose-looking rock, and visualised the lowering process in preparation. I cleared a few small, loose stones from the quarry edge and continued to visualise leaning back on the rope, feeling it go tight, then stepping out over the edge. My visualisation continued right back to the anchor and to the rope running through the belay plate. After a quick 'All OK?' I performed a final check as I followed the visualisation process to the last detail, and it was then that I noticed that the rope, although it looked like it had been threaded through the belay plate, in fact had not been. I was seconds away from leaning back on the rope, and had I done so, I would have fallen 40 m onto jagged slate boulders and in all likelihood would not be sat writing this.

Ironically, I can still clearly recall the visualisation process that took place that day, and can actually almost physically feel myself leaning back on the rope and falling. It still chills me to my core.

Visualisation requires practice and preferably specific training. In a work situation, the visualisation process is helped by first taking an overview. Even if you're directly involved in the process, you can still take an overview, a look at everything that's happening in the task, and it's crucial that someone – and preferably more than one person – take on this role. Taking an overview need not distract you from your own task; in fact, in my experience, it sharpens that up, too. Some companies employ health and safety 'spotters' to perform this role.

Task and location knowledge is very useful, too. Although a person with no or rudimentary knowledge of a task may be able to visualise a wide range of hazards and sometimes bring a fresh approach, he or she will be disadvantaged by not being able to foresee the task components in the same way that an experienced person could. It helps to know the task really well and to understand the problems that can occur.

But even this is not enough, because you also have to be prepared to act if you see something that's not right, and this can be very difficult. As a new worker in an organisation, or as a younger person with limited experience, it can be quite daunting to point out a potential error to more experienced colleagues. This is where organisational culture is so important, and in particular the creation of an egalitarian health and safety culture in which anyone can comment freely and equally on an issue connected with safety. This sort of culture has to be created on a 'bounce back', where it starts from the top, works down to everyone at the sharp end of the job, then comes back again. Too often, messages disseminate from the top down, but tend to stay there.

CONSTANT VIGILANCE AND ASSESSMENT

The concept of constant vigilance (CV) is very simple, and it's one I first really engaged with when analyzing the management of primary school groups on activity sessions. It became clear very quickly that in order to run these sessions safely, someone had to keep an eye on everything, all of the time: in other words, be constantly vigilant. I expanded this concept to working with other groups and instilled it into the staff working with me. CV is demanding, but not difficult. It's a state of mind that runs parallel with and is integral to other aspects of FTHI. In the workplace, people who exhibit CV are aware of much more than what's happening immediately in front of them. Ideally, all supervisory staff need to be in a constant state of CV. Whilst the focus of a worker performing a task is very much zoomed in on the task itself, the focus of a supervisor pulls out to a wide angle and necessitates a combined approach with visualisation.

Constant assessment is really another way of looking at dynamic risk assessment. The difference is that constant assessment is a rolling dialogue that is in a constant state of flux, whereas dynamic risk assessments tend to be performed in stages when things exhibit more significant change. In order to perform constant assessment in this context, it's essential to have secure process/task knowledge, combined with an acute awareness of changing circumstances. In practical terms, constant assessment needs to be backed up with appropriate action, which, by the very nature of this type of assessment, is often urgent.

Whilst working with groups climbing and abseiling in Langdale, the Lake District weather conditions were not always kind, and showery conditions were very common. Even in the summer, heavy showers can bring significantly lower temperatures, gusty winds and of course an extremely fast change in conditions underfoot and on the rock, all of which impact safety significantly . As a matter of course, our process of constant vigilance and assessment would take into account the changing weather conditions. We didn't wait for a squally shower of rain to hit us, we watched and observed the local conditions and anticipated the onset of rain. In most cases, our vigilance alerted us to the proximity of rain and our assessment of the situation allowed us to get people down from their climbs or abseils and to a sheltered spot before the rain arrived. We also used the process of visualisation to assess the likely conditions that would arrive with the shower, taking into account wind direction and strength and the intensity of the rainfall. This in turn enabled us to position groups to take maximum advantage of natural shelter.

PERCEIVING RISK ACCURATELY

You can read more about risk perception in Chapter 8, so suffice it to say here that an accurate appraisal of risk is integral to FTHI, and it's crucial to be aware of the factors that can lead to an inaccurate perception of risk.

So what can managers do in order to instil the principles of FTHI?

- First, managers have to accept and buy into the idea that it is possible to take current workforce observation and anticipation skills in relation to health and safety to a new level through the concept of FTHI. The implementation of FTHI may require cultural change and additional training, so there's a time and cost implication.
- Provide training for management and supervisors, then train workforce.
- Ensure managers/supervisors spend more time with younger and less experienced staff, encouraging them to observe and look for potential issues.
- Use experienced staff to train less experienced staff and draw on their experience to build collective knowledge of known hazards.
- Ensure staff understand the factors that may affect their perception of risk.
- Train staff to use a trigger that puts them into FTHI mode at the start of the day and following each break. This is an essential component of the FTHI process.
- Try running short, competitive exercises in FTHI by asking a group to analyse a particular aspect of work and use FTHI methods to individually compile a list of static hazards along with an assessment of how changes in circumstance would affect them.
- Ensure staff are aware of the need for constant vigilance and the link this has with rolling, constant assessment and the role of dynamic risk assessments.
- Use reward systems to encourage staff to identify, report and intervene in issues of safety.
- Use team-building events to learn about and establish the principles of FTHI, build relationships and create an environment that makes it easier for people to intervene.
- Stress that identification of a hazard is only part of the process and that taking relevant action is critical. Help workforce to understand what relevant action is and how to take it.

To sum up FTHI, it's a combination of a state of deliberate, heightened awareness with advanced visualisation and observational skills. It encourages people to look at managing risk in a different, free-thinking way in which hierarchies become irrelevant and the focus is on the team rather than the individual. FTHI also encompasses the principle of constant vigilance and assessment and the requirement to take immediate and appropriate action where required. FTHI mode needs to be switched on at the start of any work session, so the use of a 'trigger' to activate this heightened mental state is vital.

12 Dealing with Change

In my health and safety survey, when asked to state the most likely reasons for the occurrence of workplace accidents, 71% of respondents cited changing circumstances as the most likely reason. Change is inevitable and is often random in nature. A situation can change very quickly or it can change almost imperceptibly through increments (marginal change). However we look at change, its inevitability leads to the equally inevitable conclusion that we will have to deal with it throughout our lives, including changing situations at work.

Change has a sort of inconsistent constancy, and you'd think that we'd feel comfortable with it, as we have to face it so often. Yet many people fear change and struggle to cope with it.

So what is change? The nature of change as we would like to define it in relation to the management of safety in tasks at work is:

An act, event or process through which things become different.

Thinking about this for a moment, as soon as change takes place while we're performing a task at work, and things become different as per the definition, we will probably have to act differently to some extent in the way that we carry out or manage the task. Though it's possible that change may not affect the task, more often than not, changing conditions will require us to act in some way as a response. If change takes place quickly, it is easy to observe and we should be able to change our behaviour in response to that change. Where change is creeping and almost immeasurable in the short term (marginal change), our response to that change may be delayed or lost as we fail to recognise it until a much later stage. Most commonly, though, change will be sudden – for example, a change in the weather, a delivery may be delayed or arrive early, a piece of machinery breaks down or a colleague is absent.

This is the point at which something of a clash can occur between an approach that relies heavily on structure, procedures, training and systems and the type of free-thinking attitude that allows people to move outside the rigidity of that type of system and react to change in a positive, safety-enhancing manner. This is in no way a criticism of that type of approach, as having effective systems and procedures in place together with comprehensive training is vital to managing safety at work. Additionally, a good systems-based approach will doubtless cover many of the potential changes in circumstance that could be expected. However, if personnel are too heavily reliant on a procedural approach and something changes in a way that places them outside the conventional structure of safety management in their job, it's vital that they be able to supplement this with a free-thinking approach.

Many accidents and incidents are caused by a combination of changing circumstances, the results of which may be expected or unexpected. The random nature of change, and in particular multicircumstantial change, may present us with situations that are new to us that may continue to change and develop and that cannot be addressed from within the normal procedural channels.

Changing circumstances may, by the same token, provide much more clarity in terms of future required actions and may clearly highlight a person's lack of knowledge, experience or judgment, a trait seen many times in outdoor activity incidents. Used positively, these experiences can provide vital lessons.

THE GLENRIDDING BECK TRAGEDY

One of the most unfortunate accidents in recent years in the Lake District involved the death of a child on a school trip following an incident that occurred while the group he was a part of was undertaking a 'plunge pool' activity. This involved jumping into a pool in a river (Glenridding Beck) from a height of about 4 m, then swimming to an exit point.

The school and the teacher in charge had done the activity before successfully and, based on this experience, decided to undertake it again.

However, the conditions in fast-flowing mountain rivers such as this can change rapidly over a short period, and a shallow, benign stream can quickly become an icily cold torrent. Conditions can also change due to events some distance away; for example, heavy rain farther into the mountains might not be obvious as you move farther away from them, but the increase in rainfall upstream can lead to key changes further downstream.

There are a significant number of variables that can in turn lead to a number of key hazards changing.

- Water depth
- Water flow rate and the development of currents
- Temperature of water
- Air temperature, especially in a microclimate near the water
- The proximity of rock features to the water surface

Group leaders cannot rely on the fact that the activity worked well last time they undertook it, as almost certainly the conditions will not be identical. Simply having performed a task or activity before, even a number of times, does not make that person competent. Many outdoor activity incidents have taken place when a well-intentioned leader carries out the same activity but in different conditions, without having experienced those conditions him- or herself. The leader is therefore not able to determine the effect on the group or activity that those changing conditions may have.

The conditions on the day for the plunge pool activity mentioned above were very different from the previous times the teacher in charge had undertaken it. The water levels were significantly higher, the water flow significantly faster and the temperature both in and out of the water significantly lower. It was a recipe for disaster and I can recall very clearly, on hearing about the incident later that day, my utter amazement that anyone had attempted the activity given the dire conditions the area was suffering from at the time.

Lacking the experience to understand how the different conditions on the day could affect the activity, the teacher in charge took the decision to go ahead and the tragedy unfolded. After jumping into the water, the casualty was seen to be panicking and unable to exit the pool. The leader of the group entered the water to help him, but was soon incapacitated by the cold water and had to get out. The child's mother also entered the water in a rescue attempt, but she was also quickly affected by the water temperature and was in turn rescued by another pupil, who was also affected by the cold. In fact, both of them were airlifted to hospital and treated for hypothermia. The casualty was washed from the pool and later pulled out of the river 150 yards or so downstream, but was sadly pronounced dead.

Looking at the way the tragedy unfolded, you can see how there could easily have been more than one fatality, as is so often the case in water-related incidents where attempts are made to rescue a struggling party.

Perhaps the saddest part of this incident is how easily avoidable it was. If you look at the Ease of Avoidance Axis chart (see page 36), you will see that the incident resulted in the most serious consequence, loss of life, yet was very easily avoidable had the right decision-making process been used. Human error in the form of making the wrong decision was the reason the accident took place. Having a person with appropriate knowledge and experience present to make the decision would have prevented the accident.

THE IMPORTANCE OF PERFORMING THE CORRECT RISK ASSESSMENTS

The HSE report following this incident highlighted the necessity of having three levels of risk assessment for the activity.

Firstly, a generic risk assessment should cover the risks inherent in the activity. This could, for example, include drowning, becoming hypothermic due to cold water or striking a rock hidden in the pool.

Secondly, a site-specific risk assessment is also crucial. I used this type of risk assessment extensively for the activities I ran. It covers specific risks associated with carrying out the activity at that site. It might mention that a certain exit point from the pool should not be used due to sharp rocks, or that the approach to the jumping point should be from a certain direction to avoid a steep drop. It should also warn of the rapidly changing nature of local conditions due to the local mountain environment.

Thirdly, and crucially for many tasks and activities, a dynamic risk assessment should take place continuously during the life of the task. This takes into account the many changing circumstances that can affect the safe running of the task or activity. For example, rapidly changing weather conditions or changes in the mental or physical condition of the group would need to be assessed continuously by someone with the experience and knowledge required to make decisions based on those changes.

Had someone with the right level of experience performed a dynamic risk assessment on the day of the activity, that person would have noted the strength of the water flow and the low temperatures both in and out of the water. However, even after performing a dynamic risk assessment, the person responsible has to then make a decision as to whether to continue with the activity. I suspect strongly that the teacher in charge on the day in question did notice that the water flow was strong and that the temperatures were low, and in some format did carry out a mental version of the dynamic risk assessment. However, and this is a really crucial point, he did not have the depth of experience and judgment required to realise that conditions on the day were unsuitable for the activity. An experienced outdoor instructor would have had that depth of experience and would have acted differently in the same circumstances. The ability to carry out a dynamic risk assessment in the event of changing circumstances needs to be matched by that person's experience, judgment and decision-making ability. Only then can the correct decision be made.

FIXED CRITERIA

It's worth pointing out that the HSE report mentioned that it would be desirable to have some sort of fixed and measurable criteria for the activity that could be used to help the decision-making process. This is a very good idea in principle, but in practice can be very difficult to achieve. For example, it might be possible to measure water and air temperature, but alone, these figures are only very broad indicators. Other factors such as water flow and the effects of wind, rain and sun on the ambient conditions are much harder to measure in absolute terms. Other variables include the fitness levels, age, experience and equipment used by the participants. An experienced outdoor instructor would have put together all of these variables in making the decision, and although it is possible that one single factor such as water temperature might override all other considerations, it's more likely that all the factors mentioned above would be incorporated into the decision-making process.

Here's a list of the criteria that should have been used to make a decision about whether to run the plunge pool activity:

Water temperature: Easy and quick to measure with a thermometer on the day. No information available at the time indicating a cut-off temperature below which the activity should be cancelled. Most outdoor instructors would set this factor against other factors such as equipment (wetsuits?), experience of the group and current weather conditions (sun/ wind, etc.).

Air temperature: Easy and quick to measure using a thermometer on the day. Only tells part of the story, as wind chill, rainfall or sunshine can make huge differences on the effects of a particular temperature.

Weather: A crucial part of the decision-making process and linked with water temperature. The participants would feel very different exiting water of a given temperature into a warm, sunny environment than they would into a windy and wet one. Weather conditions at the start of the activity would need to be set alongside the local weather forecast, and the location of the site would necessitate an understanding of mountain microclimates.

Water flow rate: Difficult to measure. Only extensive experience of water activities in different conditions could allow an instructor to make an accurate judgment on whether the water flow would present a hazard.

Group age: Children are generally less resilient to cold and wet than adults; it's a factor that has to be taken into account.

Group experience: Has the group undertaken this or a similar activity before, and how many times? Can they all swim, and have they had experience with swimming in the conditions present on the day?

Mental and physical condition of group: Are the group members physically fit and mentally strong?

Group equipment: Are the group members wearing wetsuits and buoyancy aids? Helmets?

Rescue equipment available: What rescue equipment is available on site – buoyancy aid? Rope?

Experience of leaders: Have the leaders experienced this activity in similar conditions before and are they competent both to undertake the activity confidently and to rescue anyone in difficulty?

Ease of access to pool: Do the conditions on the day allow easy access and egress?

Depth of pool: Is the pool deep enough for the activity to take place? Could the depth of the pool create problems for the participants or for anyone attempting a rescue?

Presence of underwater hazards, for example, rocks: A check has to be made to ensure that there are no rocks beneath the water that participants could accidentally strike.

Going back to the theme of managing change, some of the factors mentioned above are fixed at the beginning of the activity, for example, the age of the participants and their experience levels. Other factors may change during the course of the activity; these include weather conditions, air temperature, water temperature, water flow rate and the physical and mental condition of the group.

All the meetings, child protection policies, pretrip planning and risk assessments written up at school prior to the trip are rendered utterly useless without the right person making the correct judgment on the day.

Gorge scrambling and other activities that combine water with scrambling or climbing present a complex mix of ever-changing risks.

THE STAINFORTH BECK TRAGEDY

An incident with a number of parallels took place in Stainforth Beck in the Yorkshire Dales in 2000. A school party decided to undertake a river walk – wading upstream against the current. Though one of the teachers in charge had some outdoor experience, they had no experience of instructing outdoor activities and did not follow the three-stage risk assessment process outlined above. Poor conditions on recent days and on the day of the accident led to the water level in the river rising sufficiently to wash two members of the group off their feet and downriver to their deaths.

As in the Glenridding Beck incident described earlier, the staff in charge of the children did not have the levels of experience required to make an informed judgment as to whether the activity was safe. They did not even realise that Leeds City Council had produced their own guidelines on managing the safety of children undertaking outdoor activities. The trail of culpability therefore goes right back through the school governors and head teacher to the Council's Director of Education.

Had all the guidelines been seen by the teachers in charge, one wonders if it would have made any difference on the day, as just like in the Glenridding Beck tragedy, those in charge clearly lacked the experience required to make a correct judgment.

Both of the cases mentioned above involved change. Common to both tragedies was the difference between conditions experienced in earlier visits or site recces and the conditions experienced on the day. Also common to both was the fact that the conditions locally on the day added to the level of disparity between previous conditions and those encountered at the time of the accidents.

There are clear lessons to be learned here both for the world of outdoor activities and for those working in, or managing, any job that is exposed to variable or changing conditions.

1. The fact that people may have undertaken an activity or task before does not mean that they are competent to undertake the same activity or task in different conditions.
2. The production of paperwork, systems, policies and procedures has to be coupled with staff who understand how to assess change and how to respond to it.
3. Staff should be trained in making dynamic risk assessments and free-thinking hazard identification. They should also be trained in how to manage the decision-making process that follows.
4. Staff undertaking dynamic risk assessments should have sufficient knowledge and experience of the task to be able to make informed judgments and decisions.

There are many common, everyday situations that require us to change the way in which we carry out a task.

I'm driving up the M6 on a fine, dry day, but as I approach Birmingham, it starts to rain very hard and traffic becomes heavier. As an experienced driver, I respond to this by putting my headlights on; slowing down to what I judge to be an appropriate, lower speed and increasing the distance between myself and the car in front. I might also make sure my wipers are on full speed, ensure the rear wiper is on and adjust the heater controls to ensure clear windows. I then continue to monitor the road conditions to watch out for excessive surface water and the risk of aquaplaning, and I monitor the position and actions of other traffic and maintain a speed and distance appropriate to the conditions. I'm constantly undertaking a dynamic risk assessment (constant vigilance and constant assessment), taking into account all the factors mentioned and more besides. The logical conclusion of this is that if conditions get worse and become too dangerous to drive in, I would respond by slowing down further and pulling in at the next service station or exit, parking up and waiting for the conditions to improve. Should the rain ease and driving conditions improve, I would increase my speed once more and resume my driving mode for normal road conditions.

All well and good, but as you will know if you've driven much on British motorways, not everyone will perform a rolling dynamic risk assessment and put the same set of control measures in place. Some drivers will not reduce their speed unless forced to by other traffic queuing, and others will continue to drive close behind other motorists regardless of the surface conditions and visibility. Some will drive without lights and, sadly, some will be involved in accidents.

It's hard to believe that so many drivers are prepared to ignore the increased levels of risk presented to them in a changing situation like that described, yet they do. So why do people fail to respond to changing conditions?

- *Illusion of safety*: Many people genuinely believe that they are safer than they really are when driving. The relative speed differential between cars travelling on the motorway is low, and the driver's brain focuses on this rather than the actual speed at which all the vehicles are travelling, providing an illusion of security. Drive at 100 mph all day alongside cars travelling at 110 mph, and you'll believe you're driving slowly. We are naturally fearful of heights, but because we cannot judge and rationalise speed in the same way, we show it a lack of respect. In a work situation, performing a hazardous task alongside other people performing similar tasks can reduce the perception of danger and create the same illusion of safety.
- *Dehumanisation/desensitisation*: Cars and their drivers merge as one to simply become an object to other drivers, and objects get less respect than a person would. Desensitisation is a technique used to help cure phobias, but it also refers to the process in which a person has a diminished emotional response to a particular issue through frequent exposure to it. A worker frequently exposed to a task with specific risks can become desensitised to those risks over time, a change in behaviour that can lead to overconfidence and both a specific and more general lack of regard for safety. This is especially true in situations in which a task with high levels of risk is undertaken repeatedly without incident, and even more so when others are also observed undertaking the same task without incident.
- *The Dunning-Kruger effect*: This is a cognitive bias of illusory superiority arising from the fact that 'low-ability' people are unable to recognise their own incompetence and assess their ability as greater than it actually is. Conversely, people of 'high ability' presume that tasks that they find easy to perform are also easy for other people to perform. In other words, one group of drivers thinks that they are considerably better at driving than they are, while another group of good drivers thinks that everyone else should be able to drive as well as they do. This effect has clear parallels in work situations, too. In my view, highly competent people expecting others to be able to perform at the same level as a matter of course create just as many problems as people of low ability who believe they are more competent than they are.
- *Ignorance and lack of respect*: Unfortunately, many errors are made through ignorance, a lack of thought or lack of respect for others. In the driving example mentioned, ignorance of the effects of conditions on a combination of braking distance and car stability is commonplace. If, as drivers we really did have some respect for each other, we would collectively be less likely to tailgate, speed, undertake and perform all the other dangerous manoeuvres we see daily on our roads. In a work environment, we expect training and knowledge to reduce levels of ignorance, but having thought and respect for others are attributes that are more difficult to learn.

- *Lack of training*: People accumulate knowledge and skills through a wide range of scenarios ranging from formal training to experiential learning. With little or no formal training outside a conventional driving test, learner drivers may be suddenly exposed to extremely difficult and dangerous driving conditions on busy motorways. It's unfair on them and other road users to expect them to cope with these sorts of situations without appropriate training, yet it's only very recently that the idea of compulsory motorway training for learner drivers has surfaced. This is a classic example of marginal change and our failure to respond to it. Over a period of 10 or 20 years, our motorways have become busier and busier. Driving habits have also changed, cars have become faster and quieter and to some extent the conditions in which we drive have also changed. Our societal attitude towards learning to drive has not changed very much in response, yet as individuals, we know that it should have. The provision of effective workplace training in the skills required to perform a task is now commonplace and generally performed to a high level. However, training in the context of reducing accident and incident rates through a reduction in human error and the management of the human component of health and safety has lagged behind task-based training, and this imbalance needs correcting.

- *Habit*: Actions repeated many times in similar contexts such as time and place become habits. Many of us will have been accused at one time or another of having a bad habit, while the good ones rarely get mentioned! Driving too close to another vehicle or exceeding the speed limit are examples of bad habits, while checking the rear-view and side mirrors frequently might be described as a good habit. Research indicates that there is a very clear association between a habit and the context in which it takes place. The context could be a set of environmental conditions, a time or a place. Many tasks that are performed at work regularly are habit forming, but when a situation changes, those habits could be lost due to contextual changes. Therefore, if an operative normally exhibits a certain reaction to a given set of circumstances and the context of those circumstances is changed, the person's response could be different. Established habits can also create problems when people continue to use the same habitual response to a problem they have encountered in different, but parallel, circumstances. I've mentioned elsewhere in this book the incident involving skier and paraglider Gustav Fischnaller, who, through reverting to habit, tried to use a paragliding technique to deflate an Upski sail, resulting in a serious accident.

MANAGING CHANGE: MANAGEMENT ACTIONS

The first clear action is to recognise that changing circumstances are a contributor to the occurrence of accidents and incidents. Although changing circumstances can affect any workplace situation, they are particularly likely to be present in industries such as construction, agriculture, forestry and outdoor adventure, where everything from weather conditions to late deliveries can effect meaningful change.

The second is to ensure that as far as possible, task training, procedures and systems are designed to take into account changing circumstances.

The remaining actions are those that help a workforce to deal with changing conditions that present problems and risks outside those that can be identified within a procedural approach.

- Ensure the workforce understands the relationship between changing conditions in the workplace and the incidence of accidents and injuries.
- Provide training in free-thinking hazard identification and stress the need for constant vigilance and constant assessment.
- Train workforce to carry out dynamic risk assessments and create a clear understanding of when to use them, stressing the need to start in response to the first change and the need to continuously update.
- Ensure the workforce understands the decision-making process that follows the completion of dynamic risk assessments.
- Ensure staff understand the need to assess change constantly and be aware of the importance of combinations of changing circumstances.
- Encourage staff to examine and be aware of actions performed through habit and to guard against habit-based error in changing circumstances.
- Make staff aware of the danger of transferring work methods to similar, parallel tasks.
- Ensure staff understand the concept of marginal change and are aware of how to recognise it and act accordingly through monitor/review techniques.
- Change can throw up deficiencies in knowledge or skill. Encourage staff to feel free to admit to weaknesses in knowledge and skill areas and make it not just acceptable to do so, but a really positive thing.
- Use external sources to revisit risk assessments periodically in conjunction with the workforce to help balance workforce views on safety in relation to tasks that are carried out regularly and in which risk desensitisation could have taken place.
- Ensure that even experienced staff understand that changing circumstances can produce situations in which previous knowledge and experience may be insufficient to make informed and correct decisions.
- Create a culture in which workforce response to changing circumstances is cautious and considered, yet positive and proactive and is valued and shared.
- Ensure the workforce understand that when circumstances change during a task, it is acceptable for the task to be halted until either the circumstances revert to normal or sufficient measures have been put in place to enable it to continue safely. This goes back to one of the fundamental principles of the health and safety movement, which is that health and safety are not a reason to simply stop doing something or avoid it, it's a movement that is best described as enabling, offering ways of continuing to undertake tasks that carry risks by taking appropriate measures to reduce those risks to an acceptable level.

13 Nutrition

As I write this, the UK National Health Service (NHS) is in crisis, unable to cope with the demands placed upon it during the winter months, with cases of flu inevitably making things worse. The main concern, though, is the prevalence of preventable disease in an ageing population, coupled with concerns about the diet and exercise habits of children. Obesity, diabetes and heart disease are particularly linked with nutrition and sedentary lifestyles, while many other conditions and illnesses can be linked with these factors. Two-thirds of adults are considered to be obese, along with one-third of children.

Exercise and diet are currently high on the political and social agenda, and rightly so. Many commentators are suggesting that the health problems faced in the United Kingdom need to be tackled by people taking individual responsibility. I agree with that wholeheartedly, but realistically people often need help in order to change, and the workplace can provide part of that solution through training, information and direct support.

In March 2018, public health officials called on food sellers and manufacturers to cut calories in their products by 20% by 2024, with big businesses that fail to make progress set to be named and shamed. Public Health England (PHE) estimates this could slash costs to the NHS by £4.5bn over 25 years and prevent more than 35,000 premature deaths, with another £4.48bn saved in social care costs.

'We have more obese children in England than ever before – we have one in five children arriving in primary school already obese or overweight and one in three leaving primary school obese or overweight', said Dr Alison Tedstone, chief nutritionist at PHE, adding that more than 60% of adults are also too heavy. 'It is the norm now', she said. Tedstone also said that the cost to the NHS was substantial, with £6.6bn spent every year on obesity-related illnesses.

In the workplace, the advantages of a healthy diet are clear, and offer both short- and long-term benefits.

- A good diet helps people to concentrate better and provides appropriate energy release.
- A good diet leaves people feeling less tired and more energized.
- Avoiding certain foods and drinks prevents a see-saw effect in concentration and energy levels.
- An appropriate diet controls weight, which helps people feel healthier.
- The correct foods at the right time aid muscle development, healing and recovery – important for workers in physical jobs.
- The right foods can bolster your immune system and leave you less prone to illness.

- An appropriate diet can reduce the chances of long-term issues with obesity, diabetes and heart disease, for example. This has clear benefits to both employer and employee.
- Workers eating the right foods will do a better and safer job.

Obesity trends in men and women in the United Kingdom between 1993 and 2010 show an almost doubling in the prevalence of obesity in a 17-year period.

COFFEE OR CHOCOLATE, ANYONE?

As a precursor to the way that nutrition can help us reduce incidents of accident and injury in the workplace, just take a moment to consider the side effects of the two key ingredients in energy drinks – caffeine and sugar.

Did you know that a 500 mL can of Red Bull contains 13 teaspoons of sugar and the caffeine equivalent of two cups of coffee? Or that between 2006 and 2015, sales of energy drinks rose by 185% to a consumption level of 672 million litres in the United Kingdom? That's a lot of sugar and a lot of caffeine!

Educationalists have been warning for some years that energy drinks consumed by pupils are having severe effects on behaviour and learning in both primary and secondary schools. Energy drinks are cheap and readily available – you only have to pop into your local supermarket just before school starts to see just how popular these drinks are. If they're harmful to schoolchildren, what effect could they be having on a workforce that in parts consumes equally vast volumes?

Caffeine is the most common energy drink ingredient and one of the most widely consumed substances in the world, as can be witnessed by the seemingly exponential growth of coffee shops in the United Kingdom. Caffeine is a drug. People enjoy drinking it in all its forms, and they enjoy the feeling of alertness it provides. However, there's an inevitable downside.

Caffeine tolerance varies between individuals, but for most people, a dose of over 400 mg per day may produce initial symptoms such as restlessness, increased heartbeat and insomnia. To put that in perspective, a cup of coffee normally contains between 60 and 100 mg of caffeine, but this can be up to 200 mg. An energy drink generally contains between 70 and 240 mg of caffeine. A couple of coffees and a couple of shots of your favourite energy drink could easily put you past the 400 mg mark. Higher dosages can lead to:

- Increased blood pressure
- Heart palpitations
- Anxiety and panic attacks
- Diarrhoea
- Increased urination
- Dizziness, irritability, nausea, nervousness
- Allergic reactions including rash, itching, difficulty breathing, tightness in the chest, swelling of the facial area, shakiness, trouble sleeping, vomiting
- Headaches and severe fatigue can occur during withdrawal
- See-saw levels of alertness/tiredness

Sugar is usually found in high levels in most energy drinks. There's enough information in the public domain to make sure everyone understands the hazards associated with excessive sugar consumption, that is:

- Obesity
- Tooth decay
- Increased risk of Type 2 diabetes
- Blood sugar and insulin spikes, which later result in a 'crash-like' feeling
- Sugar is also somewhat addictive

NUTRITION AND WORK

All well and good, but how does this affect us at work?

Quite simply, what we eat and drink affects, for example, our energy, alertness and concentration levels and how tired we feel.

Considering that contributory factors to many accidents and incidents are lack of concentration and tiredness, the importance of taking nutrition into account when you're working is crystal clear.

As far as this example is concerned, the quick energy and alertness boost released by a slug of coffee or energy drink may satisfy us in the short run, but the increased blood sugar levels quickly dip, often resulting in us feeling more tired than we were originally. The caffeine component, too, provides a short-term boost, but, in a similar manner, its effects recede, leaving us feeling tired once more.

Whilst we are dealing here with how nutrition affects workplace performance and its subsequent influence on health and safety, there is inevitably a wider issue associated with health and lifestyle that is inextricably linked. If we can educate and influence our workforce regarding the benefits of workplace-effective nutrition, we are likely to see long-term benefits of a wider nature, such as better levels of health and fitness, a lowering in the rate of illness in later life and a reduction in the demand placed on the NHS and other health resources.

THE BIGGER PICTURE

Having said this, the reality is that, as a society, we have been trying for many years to help and guide people towards a better and healthier diet, but it has proved extremely difficult to change people's eating habits, and some radical approaches will be required alongside the current war of attrition waged by many health organisations.

Stepping back a little and looking at the bigger picture regarding nutrition, it's hard not to feel that as a society we have been commercially manoeuvred towards a poorer diet, and that without some concerted action, today's youth will experience health problems at an unprecedented level. Every garage you go into piles up the chocolate and sweets in front of you as you approach the till, and most display coffee machines. Many shops and supermarkets stock vast quantities of high-sugar food and market it aggressively. Advertising and commercial sponsorship pushes unhealthy products at us. Last night on TV, one of the prime-time ads boasted that your evening was 'sorted' if you took up their offer of chips, pizza, ice cream and coke. In my local town, if you go into the supermarket at 8.30 a.m., you'll see dozens of children piling the sweets and sugary drinks onto the checkouts. It's actually quite horrifying to observe.

You could maybe make some comparisons here with the history of smoking and the sale of tobacco-related products. There was a significant lag time between the understanding of the terrible effects smoking can have on our health and any serious action to reduce the problem. The commercial clout and slick advertising of the tobacco companies inevitably affected this, but so too did the fallibility and entrenched attitudes of a generation. Similarly, the increasing concern regarding the effects on our health of a poor diet are very slowly starting to change our attitudes, but our concern has not yet led to changes in the way that unhealthy products are marketed, or to the sort of social unacceptability that affected smoking.

So what greater encouragement could we need to try to influence the way our workforces eat and drink? We can reduce accidents and incidents at work; we can help our workforce to live longer, healthier, happier lives and we can reduce the burden on the health service, freeing up money and reducing patient waiting times. Big goals require big effort, and it's here that you can see the principles of marginal gain at work. We're never going to persuade everyone to adopt a healthier diet, but every person we influence makes a difference. Every person who changes the way he or she eats in a positive way makes a difference. Every accident that doesn't happen because someone is concentrating more effectively and every illness that doesn't develop due to dietary change eventually add up to make a noticeable impact.

I strongly believe that of all the measures outlined in this book, changing the way that people eat and drink can have the most powerful and immediate impact, together with the potential for long-term, positive change. The short-term benefits can be backed up by evidence from the construction process of London 2012. Construction workers were provided with subsidized porridge breakfasts, which attracted a big take up and a reported decrease in the number of accidents, incidents and near misses.

As further evidence of sorts, I elected to undertake my own experiment, to assess the effects of two radically different diets. For two weeks, I deliberately ate what nutritionists would call a poor diet. I loaded up on biscuits and chocolate, ate lots of carbs, drank some beer in the evenings with crisps and nuts, and generally ate too much. The result? I felt awful. I was tired, particularly in the afternoons, I felt like my energy levels were significantly lower than normal and in an overall sense, I just felt below par. Interestingly, old injuries ached and I felt physically dulled. The next two weeks saw a complete change of diet. I ate no chocolate, biscuits or other sugary food, drank a lot of water, ate plenty of low-fat protein, veg and fruit, and cut down on carbs, only eating a small amount of wholegrain bread. After a week, I started to feel progressively different. My alertness levels were increased, and I felt less tired and had lots of energy. I felt fitter, psychologically more positive, and the aches and pains disappeared. The difference was truly remarkable; you should try it for yourself because it could provide all the encouragement you need to inspire your workforce to join you.

COMMON PROBLEMS

Here are some of the most common problems associated with diet and nutrition that can impact safety in the workplace:

- Energy drinks with a high sugar content lead to excessive blood sugar level fluctuation and spikes and troughs in how tired we feel.
- The caffeine component of energy drinks and coffee has a similar effect, leading to highs and lows.
- Eating sugary snacks such as bars of chocolate also leads to blood sugar level fluctuation.
- How often you eat during the day can influence tiredness and concentration levels. Leave it too long between meals and you may struggle to concentrate and feel tired. Some people are significantly more susceptible to this than others.
- The type of food you eat at breaks can seriously affect how you feel afterwards. Too much sugar and you may experience sugar level fluctuation, too much fatty food and you may feel tired due to your digestive system's energy consumption. Eating simple carbohydrates has been linked to increased fatigue, whereas eating complex carbohydrates reduces fatigue by sustaining blood sugars – without them, the body loses steam and you become tired.
- After increased levels of physical exertion, eating the right balance of carbohydrates and protein within 45 minutes of completion maximises recovery; failure to do this can leave you tired and aching.
- A poor diet can lead to a shortage of key vitamins and trace elements. Anaemia caused by a lack of iron is a common and well-known issue.
- Dehydration reduces our physical capability by up to 20%. I came across a men's toilet block on a building site that had a chart showing how the colour of urine linked with stages of dehydration – the broad principle being that the darker the urine, the more dehydrated you were. A useful reminder, but in some ways a reflection of failure, as a nutritionally aware and educated workforce would consider this as a matter of course.

Many of the factors that contribute towards human error are a direct result of, or exacerbated by, what we eat, how much we eat and when we eat. As an example, it has been established beyond doubt that working extended hours results in a significantly higher exposure to workplace accident and injury. In many cases, working extended hours is linked with a failure to consider diet and breaks appropriately, adding to the problems that working extended hours brings.

TIMING OF BREAKS

What we eat is very important; so, too, is when we eat it. Studies in Spain and the United States have shown that accident rates may be at their highest and most serious either side of average lunchtime. This factor could also be expressed as proximity of last meal or next meal rather than as a chronological statement. I believe that more rigorous study is needed to establish this link, but anecdotal and personal experience backs this theory up. One report concluded that working long hours doesn't put workers at higher risk of accident or injury only because long hours are concentrated in hazardous occupations, or because there is more time spent in dangerous environments. It went on to say that managers should consider changes in scheduling practices, job redesigns and health protection programmes for people working regular overtime and extended hours.

As we approach lunchtime, our energy levels are likely to be depleting, and our focus is shifting away from the task to the thought of lunch. After lunch, depending on what we've eaten, there are two factors that could influence the potential for an increase in accident rate. The first is due to lack of concentration in that period after lunch when someone's starting work again and switching from lunch to work mode, and FTHI is the way to deal with this. The second is that depending on what food has been eaten, an overworked digestive system can cause a period of tiredness, as energy is diverted to the digestion process. A lot of sugary food at lunchtime can lead to a short-term energy rush, but it doesn't take long for that effect to wear off and tiredness to set in.

In mountaineering and hill walking, there's always been something of a mantra about eating little and often. The theory has always been that it evens out the body's process of breaking down food, and it releases a more constant flow of energy. There's some truth in this, and factors such as evening out blood sugar levels are also important, to avoid those peaks and troughs in alertness and energy levels.

About a quarter of people who responded to my health and safety questionnaire said that a different pattern of break times would help them to stay more alert in the workplace. Many people will not have even considered the potential benefits of a different work/break pattern, and remain governed by a historical template that dictates breakfast, lunch and dinner, usually with a snack in between each.

WHY IS THIS RIGHT? SHOULD IT BE DIFFERENT?

Opinions vary, but many health and fitness experts recommend that taking six smaller meals a day provides a better and more constant release of energy and a more even blood sugar level. Based on my experiences in demanding mountaineering situations, I'd agree with this principle. Breakfast remains an important meal, and is often taken outside the work environment, though some larger-scale work projects have brought breakfast into the workplace as an important nutritional strategy.

How to implement changes to break patterns in the workplace is another matter, because there are a number of cultural and organisational obstructions. The first one is that people may not agree, and may want to stay with their three-meals-a-day routine. The second is that when co-working, it might be beneficial if breaks are at the same, or approximately the same times.

WHAT TO EAT AND WHAT NOT TO EAT AT BREAK TIME

There is clear evidence that certain foods place extra strain on the digestive system, which in turn uses extra energy that can lead to tiredness. Other foods and substances can cause energy spikes and troughs, whilst certain carbohydrates provide effective slow release energy and others don't. Some foods are additionally reputed to aid concentration levels.

Examples of What Food and Drink to Promote

Complex carbohydrates, as in wholegrain bread or rice, fish, eggs, fruit, vegetables, grains, poultry, nuts, beans, low-fat dairy products. Blueberries, salmon, beetroot, kale and spinach, avocados and dark chocolate are amongst a number of foods that nutritionists say especially help with concentration and alertness.

Examples of What Food to Avoid

Processed foods, refined carbohydrates such as white bread and white sugar, saturated and trans fats, sugary snacks, caffeine, alcohol, sugary drinks.

Don't Forget Water

Water is often underestimated as an indirect source of energy and focus, with dehydration a common cause of lack of energy and concentration. Provision of drinking water and making sure a workforce is fully aware of dehydration issues are essential. Water moves food through your intestines, helps regulate your body's temperature and helps with joint movement. It's also crucial in the production of energy molecules. Dehydration is one of the main causes of tiredness and lack of energy. If you are not well hydrated, instead of supplying you with energy, your body will focus its resources into maintaining your water balance.

Balance Nutrition to Type of Work

It's essential to take into account the type of work you're doing when considering nutrition. An example would be that although in principle we should look to eat slow-release carbohydrates for a more even release of energy, if our work is very physical and intensive, supplementing this with quick-release, high-energy foods such as raisins, bananas or chocolate at a rate of 30–60 g per hour will help to maintain blood glucose levels.

Management Strategies

People don't always take kindly to being told what to eat or drink, or perhaps more importantly what not to eat and drink. Many view it as their personal choice – a private matter over which state and employer should exercise no control. On the other hand, there's plenty of visible support available – hardly a day goes by without a mainstream TV programme on health and diet, and all other forms of media are currently equally enthusiastic about the subject. This level of media and public interest, plus the longer-term concerns aired by the NHS, at least ensure that most people will have some basic understanding of dietary and health issues, regardless of whether they have taken any action.

Managers will need to encourage, enthuse and tempt rather than coerce and insist.

As usual, your first challenge is to lead by example, and demonstrate practical and wholehearted support for schemes at work that relate to nutrition. You must get behind the concept that accident and illness rates can be positively influenced by workforce nutrition and that the long-term health benefits are well worthwhile.

- Your case will be assisted by the provision of supporting evidence and information, and you should consider bringing in a specialist in nutrition to talk to staff and present ideas and information to support your strategies.
- Using references and crossovers with sport and other activities outside the work environment that people can relate to can positively influence attitudes.
- In the London 2012 construction phase, workers were offered a subsidized porridge breakfast. This popular move led to a reduction in accidents, incidents and near misses. Where possible, the provision of subsidized food is an ideal way to kick-start the concept of improved diet. This could be done through canteens where they are available, or, in smaller workplaces, food can be brought in.
- Following on from the last point, you need to make it as easy as possible for people to eat healthily.
- Make improving nutrition a shared target with your workforce and encourage them to come up with ideas and be involved. The more a workforce is involved and takes ownership, the more likely it is to respond to the initiative.
- Establish the relationship between nutrition and a healthy long-term lifestyle and ensure your workforce understands what can be achieved in certain timescales. Promote the concept of the marginal gains approach to nutrition.
- Develop strategies around the working day to allow experimental changes to break patterns. Try shorter and more frequent breaks and remember there's no absolute pattern in this – be prepared to try different ideas. Experimentation with break patterns and nutrition is a serious business and is not to be taken lightly; you should try to make it as much fun as possible for your workforce and engage them as much as possible in decision making. A key factor to remember concerning pattern of breaks is that some people may have different metabolic rates or have more difficulty in maintaining blood sugar levels. These factors need to be taken into account – hence the requirement to involve a nutritionist for maximum benefit. Cultural and gender differences may also have an impact on break patterns and need to be taken into account where relevant.
- Crucially, you must develop ways of monitoring and reviewing changes that are made to working patterns/breaks and changes that your workforce make nutritionally.
- Encourage managed experimentation and back this up with targeted support, professional advice and subsidization.
- Ensure that as many people as possible give feedback on the effects of change. Only then is it possible to evaluate the changes and obtain worthwhile feedback that can be used to back up your strategies.

- Encourage your workforce to record and share their experiences, but establish a baseline to start with. For example, ask them to record what they eat every day for a week alongside recording how alert/tired they feel, whether they have a particular 'down time', or if they have spikes and troughs in energy levels and concentration. If you're brave enough to try and establish different work/break patterns, once again ensure that accurate feedback is obtained. In both cases, allow sufficient time for the measures to take effect.
- Ensure that workers in physical jobs that have varying levels of intensity are aware of the importance nutrition plays in recovering from high-intensity work. As most athletes know, taking in carbohydrates and protein through a combined carbohydrate/protein shake within 40 minutes or so of intensive exercise has a very positive impact on recovery.
- Encourage a reduction in caffeine consumption through coffee and energy drinks. Encourage feedback.
- Where possible, use a specialist on a long-term basis to support both management and workforce.
- Jobs vary in type and in intensity of physical output, and people vary in size, shape, culture and background and response to food. You need to be careful to ensure that these factors are taken into account when developing strategies to improve workforce nutrition, and, once again, professional help should be sought.
- Ensure adequate provision of drinking water.

Nutrition might be a complex topic, but simply eating the right food at the right times and avoiding the wrong food and drink at the wrong times keeps you more alert, healthier and less prone to accidents and injury. For a relatively small investment, there are huge gains to be made.

14 Fitness for Work and Life

Achieving good levels of physical fitness helps us at work, in our leisure activities and with our long-term health. If workers are performing physical tasks, they need to be fit enough to carry out those tasks safely – and by safely, we mean without harm to themselves or to others. Workers with relatively sedentary jobs will also benefit from physical fitness. They will feel less tired and will be able to concentrate more and enjoy a better feeling of well-being. A fit workforce should have increased resilience against damage to muscles, ligaments and tendons and be less likely to suffer acute injury from non-impact-related accidents.

A fit and healthy workforce creates a good impression and is more likely to be alert to safety issues and able to concentrate better for longer. I have no doubt that a fitter workforce is a safer workforce.

The long-term benefits of fitness include lower risks of illness such as diabetes and heart disease, the likelihood of a longer working life, greater enjoyment of leisure time and a better overall feeling of well-being and happiness.

Fifty-two-year-old Dave Birkett is a living legend of the climbing world. Since the 1980s, he has produced a steady stream of exceptionally difficult rock climbs, many of his hardest first ascents having been made on the crags of his beloved Lake District.

What makes Dave's achievements especially impressive is the fact that he's kept on working right through his climbing career as a stonemason and dry stone waller – tough jobs that would make most people welcome an evening meal and a rest in an armchair.

I spoke to Dave recently and asked him about his work and in particular about his approach to health and safety, as well as fitness. 'It gets in the way a bit to be honest', was Dave's first and in many ways unsurprising sentiment regarding health and safety. 'It just slows things down and doesn't really help matters'. Having seen Dave balance a ladder on the apex of a roof in order to climb it and gain access to another roof, while both roofs were covered in frost, his approach to health and safety didn't surprise me at all. Yet, having worked with Dave on many occasions, I also know how much care and thought go into everything he does. He might not work to a conventional risk assessment or method statement or follow any sets of procedures, but rather like his climbs, every move is carefully calculated, every step taken with precision and care. Watching Dave climb is an education. The absolute precision in every single move he makes is quite remarkable, and he's translated this approach into his working life.

The excessively high risks associated with the highest level of adventure climbing demand a meticulous approach and preparation, along with extraordinary levels of fitness and mental strength. By approaching work in the same manner, Dave reckons that he works as safely as anyone.

Beyond this lies a bigger picture and a more interesting trend, at least here in Cumbria. Dave described how one building contractor is now employing people on the basis not only of their experience and skill, but particularly on whether they undertake outdoor activities or other higher-level sport outside the work environment. The reasoning is that people who are exceptionally fit exhibit a greater work output and are less likely to have accidents. Dave agrees with this way of thinking, and when seeking help with building projects, he also looks to take on people who are extremely fit through their out-of-work interests. 'There's lots of guys working who do a decent job, but aren't very fit. I'm looking to work with people who are exceptionally fit because they get more done and they're much safer as a result of their fitness'.

I also spoke to several other climbers I know who run businesses that employ people in practical, physical work that carries risk. They shared a common belief that generally agreed with the opinion Dave expressed above, and they also shared an exemplary health and safety record. To my knowledge, no one has undertaken any research on the relationship between fitness as a result of out-of-work activity, productivity and accident rates, so at this stage, it's all anecdotal and based on personal opinion, but persuasive nonetheless.

FITNESS FOR THE WORKPLACE

Physical jobs require the workforce to exhibit a certain level of general and specific physical fitness in order for the work to be carried out effectively and safely. In many cases, the specific strengths and other physical attributes required for a particular task are developed over a period of time. Muscles, ligaments and tendons adapt and strengthen to the demands of repetitive physical movement.

Studies of the skeletal remains of mediaeval warriors reveal that the archers among them had developed distorted upper bodies that enabled them to withstand the repeated pulling of bows with extremely high draw weights. Examination of the shoulder areas revealed larger bones and significantly increased attachment points for tendons in comparison to the remains of non-archers.

Climbers develop very specific shoulder, back, forearm and finger strength over a period of time. Racing drivers have to develop extremely strong neck muscles to cope with the G forces experienced during cornering over long periods of time.

Our bodies adapt well to the requirements of specific tasks, but it's important to understand that these specific strengths take considerable time to develop. It's not just the muscles that have to adapt and strengthen, it's the vital tendons and ligaments that provide stability and attachment that need to strengthen, too and this takes longer.

Fit, healthy and young newcomers to climbing frequently suffer from ligament and tendon injuries, because these have developed less quickly than their muscles.

Some people believe that a high level of fitness makes workers less prone to injury and accident at work.

Likewise, in a work environment, both employer and employee should recognise that the strength required to perform tasks repeatedly has to be acquired over a period of time, and allowances have to be made for this.

Another consequence of performing repetitive physical tasks is the muscular imbalance created. Most muscles have an opposite – for example, the biceps have the triceps behind them – and developing one set of muscles at the expense of another parallel group can result in susceptibility to injury. It is therefore in everyone's interests to have a workforce that is both fit for the work it undertakes and also possessed of fitness in a more balanced and general sense.

TAKING THE LONG VIEW

In terms of health and fitness, it's essential to take a long-term view and herein lies one of the key problems – too many managers and workers are disassociated from this concept, partly as many of the benefits are longer term and therefore may show little demonstrable short-term gain. I'd like to refer once more to a key point I made earlier concerning marginal gain. Much of modern society expects fast outcomes. We live in a world where speed of service, delivery of goods, internet speed, travelling time and countless other aspects of modern life are seen as successful when they are fast and have immediate impact. The knock-on effect of this is that much of society now expects to see short-term, high-impact outcomes. Waiting is a thing of the past.

We can see some speedy benefits if we go to the gym regularly, visit a climbing wall or take up running. What we can't see, however, are the benefits that increased levels of health and fitness can have on our work, our personal lives and on society in general 10, 20 or 30 years ahead. This is one of the reasons targeted programmes that aim to improve workforce health and fitness can flounder after successful launches.

The short-term gains are quickly measurable. Take visiting the climbing wall as an example. After just a few visits, you will feel fitter and stronger and you'll start to do climbs with ease that you struggled with at the start. It's the same with the gym – a few weeks into a training programme, you'll feel stronger and will be able to measure the gains through the weights lifted and number of reps you can do. As time goes on, most people start to experience a plateau, a period during which the gains become less observable. This can be a difficult time, and many people drop out of their training regimens as the gains become less obvious and there's a temptation to think that having attained a certain degree of fitness, the job's done.

The statistics from research into gym membership back this up. The vast majority of people who join gyms quit during the first 6 months. Some studies reveal that between 40% and 80% of people with gym memberships don't actually go to the gym at all. That sounds like a good business to be in!

Returning to how we can establish a culture of health and fitness at work, there is a parallel between the average person going to the gym and the success of various schemes at work – things often start well and with good intentions, but often flounder as the gains lessen, as a certain level is reached that makes people feel good and as enthusiasm wanes. It can be as simple as the summer holidays getting in the way and interrupting the process.

Managers need to understand this, and when launching schemes that promote health and fitness, they need to prepare for the inevitable lull that follows some months afterwards.

HOW TO CREATE A FITTER WORKFORCE

- Management must promote and engender a culture of health and fitness and lead by example, setting a clear and stated aim of creating a fitter workforce.
- Management must establish a culture in which training for fitness at work is a normal and accepted part of work.
- Ensure that the aims of the health and fitness programme are aligned with organisational principles and goals.
- Involve the workforce as much as possible, try to pass on as much responsibility to the workforce as possible and make fitness their agenda as well as yours.

- It's essential to provide opportunities for the workforce to engage with health professionals to secure best outcomes.
- This should take the form of initial assessments, followed if at all possible by individual improvement plans.
- Design a long-term programme that will fulfil current and anticipated future needs and be prepared for interest in schemes such as this to die down after launch. Think about ways of sustaining interest, perhaps highlight successful case examples internally or promote other fitness events externally. Build workforce pride in their fitness.
- Management and the workforce should understand the marginal gains and long-term benefit theories that underpin workplace health and fitness.
- Take into account age of the workforce and cultural, gender and social differences that impact how people might improve their fitness levels.
- Look for opportunities to encourage out-of-work fitness training by providing subsidized membership (e.g., gym), provide 'have a go' sessions and so on. Don't be too narrow with ideas for helping in this respect – subsidized gym membership isn't necessarily the best option; you could consider subsidizing climbing wall membership, yoga or lots of other fitness-related activities. 'Have a go' sessions can be good for team building as well.
- Use outdoor-based activities as much as possible to take advantage of the added benefits of the outdoor environment (see the chapter 'A Connection with Nature and The Great Outdoors') and remember that you don't need to climb Everest – just walking is a great way to keep fit.
- Establish a baseline for fitness levels and review/monitor regularly. Use questionnaires as well as hard data to evaluate, and make this a long-term objective so that rates of accident, injury and near miss can be compared.
- For particularly physical jobs, consider using a physiotherapist to examine working practices and make suggestions for fitness training that complements the job.

15 Injury Reduction and Management

Injury reduction and management are close bedfellows with health and fitness, as in general terms, people who are fit and take their physical health seriously are less likely to sustain injuries. A high level of fitness can provide a useful edge in practical work, some spare capacity, if you like. It also renders you less likely to suffer from muscular, ligament and tendon damage.

Looking at the figures nationally, injuries as a result of accidents or chronic and other musculoskeletal conditions (as a result, e.g., of repetitive overuse) led to 14.5 million days being taken off work in the year 2016/2017 at an estimated cost of £5.3 billion. These are mind-boggling figures, so huge that they perhaps mask the suffering and hardship of the individuals concerned and their families and make it difficult to quantify the loading placed on healthcare and other social services.

Surely we can do better than this.

The prevention of first-instance injuries is clearly a priority, but it's equally essential to manage all injuries, whether sustained at work or away from the workplace, and fitness levels in order to produce the best possible outcomes for all concerned. *Our key aims are therefore*:

- Reduce the number of accidents that cause injuries.
- Reduce the effects that these injuries have.
- Establish early intervention strategies and manage all injuries quickly and efficiently with the aim of preventing longer-term chronic conditions.
- Identify and treat chronic conditions as early as possible, and where necessary retrain and reintegrate into the workforce.
- Examine ways of redesigning working practices and equipment in order to reduce the likelihood of injury.
- Ensure management and the workforce understand the marginal gains and long-term benefit theories that underpin workplace health and fitness.

At this point, it's worth considering just what's at stake when employees are injured at work.

First, there's the direct effect on the person involved. This could vary from short-term and mildly irritating pain through to long-term chronic illnesses that may be physically painful and mentally debilitating. Injuries can also lead to loss of earnings and other financial difficulties, coupled with a lowering of self-esteem and loss of confidence.

We then need to think about the consequences of injury on the victim's family. They may suffer financial hardship and face other difficulties as a result of the effects of the injury on the well-being of the victim.

Society suffers as well. The NHS has to fund treatment, and other government departments may have to provide short- or long-term financial support or care.

The business that the victim worked for may be adversely affected, too. There may be financial consequences and legal problems, the business may lose the essential services of a skilled and experienced worker in the short or long term and productivity may be affected.

The sheer scale of the level of absenteeism and the immensity of the subsequent costs, coupled with the potentially debilitating effects on individuals and families, make this a critically important and highly worthwhile part of our aim reduce workplace accidents and injury.

To achieve such a reduction requires both management and workforce to buy into the concept that alongside reducing the number of accidents that result in injury, improvements in the levels of health and fitness amongst employees, along with an effective early intervention and injury management programme, will benefit all parties in the longer term.

REDUCING THE NUMBER OF ACCIDENTS THAT CAUSE INJURY IN THE WORKPLACE

Achieving a reduction in the number of accidents in the workplace is one of the key aims of this book. In an overall sense, all the different strategies discussed, from FTHI to nutrition, to well-being and so on, will help to achieve a lower accident rate.

REDUCING THE EFFECTS OF ACCIDENTS THAT CAUSE INJURY

Personal Protective Equipment

One of the key factors in reducing the effects of accidents lies in the provision of excellent personal protective equipment. There have been significant developments in the design and quality of personal protective equipment (PPE) in the last 30 years in particular, a process that has been mirrored in the world of extreme adventure sport.

In skiing, the advent of excellent transceivers to help locate avalanche victims has been combined with the development of new products such as air bags that help keep a skier at or towards the top of the snow if caught up in an avalanche. Modern ski bindings won't prevent everyone from damaging knee ligaments in a fall, but increasingly effective designs reduce the likelihood.

Climbers have access to improved fall arrest systems, with equipment that absorbs energy better and is lighter and more comfortable to use. Paragliders have options for back protection and reserve chutes that just weren't available when the sport started.

Today's workforce has access to a wide range of highly effective safety equipment, and it's up to employers and workers to make sure that the best use is made of these developments. Regardless of the legal situation pertaining to the provision of PPE, it's essential for all concerned to maximise the benefits that using the best PPE can bring. This may mean looking beyond the most obvious and commonly available equipment and researching new developments. Keep up to date, and encourage your workforce to come up with ideas. Many UK soldiers have bought their own footwear rather than rely on that provided through the armed forces, recognising that to do so led to increased comfort and better performance. Don't accept second best – make sure your workforce is equipped to the highest standards of comfort, performance and safety.

Warm Up for Work

I remember watching one of those home improvement programmes recently in which the presenter made some rather sarcastic comments concerning preparation for work. His comment was along the lines of 'Health and Safety told us we had to warm up for work…' After that, his expression said it all – it clearly wasn't on the agenda. However, warming up for work makes complete sense, especially when jobs are very physical. Professional athletes and sportsmen and women don't just turn up and start a race or a game, they perform a set of carefully thought-out warm-up exercises that have two main functions – injury prevention and performance improvement. Performing warm-up exercises prepares us for physical activity by stretching tendons, muscles and ligaments and raising our cardiovascular rate. Adam Lincoln, a professional competition climber, said that the most important part of warming up for him was to combine stretching with doing a few very easy climbs, copying the movements and muscle use, but at a much lower intensity.

Perhaps the most important reason for doing a warm-up is to prevent injury during exercise; keeping the muscles warm will prevent acute injuries such as hamstring strains and will improve the reaction to overuse injuries by allowing the body to prepare steadily and safely. Warm-up sessions are also a good time to prepare mentally for what lies ahead, and this could be linked with the concept of 'crossing the line' in preparation for FTHI mode.

If you can secure advice from a fitness professional who understands the types of work involved, warm-ups can be tailored to those tasks and can provide a very beneficial start to the day. Not everyone will want to do warm-up exercises, so expect some resistance, but ask those that do warm up if it makes them feel better and use their positive messages to spur on others.

Immediate Treatment

In the event of an accident that causes injury, it's essential to have systems in place that allow the quickest possible treatment. Simple examples include having ice available for immediate treatment of muscular injuries/sprains and so on (research shows early use of ice is the most effective time) and easy access to a burns kit. Having a few first-aiders dotted around the place is great, but in my view, everyone should have some basic first-aid training that's relevant to work. HSE regulations state that minimum requirements are for a person to be appointed to take charge of the situation in the event of an incident, but this has to be related to the nature of the workplace, and most worksites will need trained first aiders. Thought should be given to the less obvious factors here such as lunch or break time cover, illness cover and holiday arrangements.

It's worth remembering what first aid is – it's the provision of the best help you can provide until more experienced help is available. For example, you might apply pressure to a wound that is bleeding excessively while waiting for a more experienced person to arrive and to apply a dressing. In many workplaces, people are afraid of intervening or providing first aid because they are scared to do so, either because they don't consider themselves qualified or because they are afraid of the consequences. Management should plan for a whole-workforce intervention strategy that increases the speed with which an injury at work is dealt with.

Early Intervention Strategies

Diagnosing and treating injuries and illness promptly through early intervention strategies is cost effective and provides many benefits to employee and employer, not to mention wider society.

Anyone involved in an accident that causes injury needs immediate medical assistance, as we all know, but beyond this, intervention is critical to ensure a speedy recovery and early return to full health and subsequently to work.

Other injuries that may have developed inside or outside work also need to be identified, and a similar process of intervention is required.

In general terms, injuries can be divided into two categories – acute and chronic.

Acute injuries are those arising immediately from a sudden trauma of some sort. Examples of acute injuries could be a blow to a limb causing a fracture; falling, tripping or slipping; a twist that damages muscle, ligament or tendon; or trapping a finger in moving parts.

Chronic injuries are long-term conditions that may arise from within or without the workplace (or a combination of the two). Examples of chronic injuries include repetitive strain injury (RSI), damage to the carpel tunnel from long periods of typing, major joint problems such as hips and knees, or damage to soft tissue that does not heal correctly and poses longer-term problems.

Acute injuries may, if not treated correctly, lead to chronic, long-term problems. This is one of the reasons a stringent policy of early intervention is so important.

Equally, chronic injuries may place a worker at higher risk of sustaining acute injuries.

There is absolutely no doubt that everyone concerned – the injured party and his or her family, the social and healthcare structure that supports the employee and the employer – benefits greatly from efficient, professional and structured early intervention. This may have to be delivered outside the NHS. Here's a case in point.

A Cumbrian manufacturer employed the services of a general practitioner (GP) and a highly experienced physiotherapist on a part-time basis to deal with medical issues that were affecting its workforce. The medical staff was able to offer treatment immediately postinjury, and for chronic conditions, they were able to diagnose the precise problems and help the workforce to return to work faster. Additionally, they were able to get to the root cause of some of the most common, recurring problems.

Workers stationed at a conveyor belt as part of the manufacturing process frequently suffered from back pain. At this point in the process, the workers had to reach to the back of the conveyor belt, which was quite wide. This involved continuous stretching and bending, which in turn led directly to the backaches described. Consultation between the manufacturers and the physiotherapist led to a modification being made to the conveyor belt that reduced the distance the workers had to reach. That simple modification stopped the incidence of back problems and resulted in better health for the workforce and a more stable and cost-effective outlook for the employers.

The company saved approximately three times the amount it spent on employing the medical professionals and saw a reduction in both short- and longer-term absences from work, along with a financial benefit from a reduction in payoffs due to the development of long-term conditions. This level of cost benefit is of a similar proportion to that identified in independent research by groups such as PricewaterhouseCoopers (PwC).

This level of intervention and support is simply not available through the NHS, which supplies an invaluable service, but is increasingly struggling to cope with the more general demands of an ageing population and organisational difficulties. The knock-on effects cast severe doubt on the system's ability to provide effective intervention and support for many people suffering from workplace injuries.

I discussed the practicality of using the traditional route through the NHS to achieve the same goals of early intervention with the physiotherapist.

> The flow in the NHS system lets people down. The system wastes time and prolongs or increases problems as people wait for appointments with GPs and subsequent appointments with appropriate specialists – which can be months ahead.
>
> She went on to say 'The faster an injury is diagnosed and treated in the correct way, the less chance there is of that person developing longer-term problems, the faster they return to work and the employer benefits financially too.'
>
> Depending on the nature of the injury and the area you work in, the NHS may not be able to provide appropriate intervention and support, so employers should look at alternative strategies for delivering fast and effective diagnosis and treatment.

PREVENTION IS BETTER THAN CURE

I mentioned earlier that changes in the design of a conveyor belt were made to reduce the distance employees had to reach – this in response to a high incidence of back problems. Although engineering professionals made the design changes, the physio working with the company catalysed the redesign. This demonstrates very clearly that a multiagency approach to the design of equipment and machinery can pay dividends.

In another example, police traffic officers were regularly complaining of backache after driving. It was found by the physio involved that the waistbelt equipment carriers they had been issued were affecting their seating position and, in many cases, this caused back pain. The carrying system was redesigned and the back problems disappeared.

The two examples above illustrate the value of using appropriate medical professionals to help find the root cause of injury and furthermore to help ensure that redesigns are carried out in a way that eliminates the problem. It's all too easy to treat people for injuries without addressing the underlying reason for them sustaining the injury in the first place.

Even better would be a situation in which working practises and procedures were scrutinised and assessed by medical professionals prior to them being adopted. This would lead to the identification and realignment of potential problems at the earliest possible stage.

MANAGEMENT ACTIONS

- Provide training and education so the workforce understands the physical requirements of their jobs and understands what they need to do in order to minimise the risk of musculoskeletal injury.
- Provide specifically targeted advice to the workforce through the appropriate medical professional, for example, physiotherapist.
- Involve medical professionals in analysing working methods and practises with the aim of preventing injury through task and equipment design and management.
- Monitor injuries, time off work and costs to evaluate intervention programmes and ensure that there are 'before' figures to make comparisons with.
- Provide workforce access to appropriate medical professionals on a regular basis with the aim of preventing injury through early, preinjury intervention and ongoing injury management. (Frequency is important – better to have a physio visit for an hour or two weekly rather than for a morning every month.)
- Actively encourage the workforce to engage with medical professionals and establish a culture that accepts this type of intervention.
- Actively encourage the workforce to take action at the onset of injury, whether chronic or acute, to enable speedy intervention.
- Look at ways of redesigning working practises and equipment in order to reduce the likelihood of injury.
- Ensure rehabilitation, return to work and absence management schemes are optimistic, person-specific and practically focussed on a positive return to work.
- Monitor intervention results.
- Take into account the age, gender and culture of your workforce and manage appropriately; for example, an older workforce may be more liable to suffer from chronic injuries.
- Intervention schemes cost both time and money, so you must be prepared for this.
- Looking after your workforce's health and providing strategies that aim to reduce incidences of accident and injury are likely to lead to an increase in staff loyalty.

16 Well-Being

A 2017 survey of 2,000 workers across the hierarchical scale in a range of industries showed that one in three people may have a well-being problem, with anxiety, stress and depression the key contributors. A quarter of those surveyed said that their employers did not take well-being seriously, and almost 40% had either taken time off work or reduced their responsibilities due to well-being issues. Eighty-three percent of respondents believed that their well-being influenced how productive they were. If well-being has that much impact on performance, it must have an equally significant impact on health and safety. Physical and mental well-being are critical links in the accident, injury and absence from work chain.

In an age in which thousands of management consultants the world over offer training and coaching to help people become successful managers, I found it fascinating to have a conversation with one of the most successful CEOs of my generation regarding what it takes to be successful at work, and there were no management-speak clichés.

You need to be secure in your relationships and family, you need to be secure financially and you need to be in good health.

When this type of advice comes from someone who has successfully managed some of the biggest construction projects in the world, it's got to be worth listening to and what he said so succinctly forms the key cornerstones of well-being.

Good level of physical health
Secure and positive mental well-being
Stable personal and family relationships
Financial security

Whilst that vision of well-being relates to successful management, it's also directly applicable to the type of well-being we refer to in relation to its effects on health and safety at work. So what exactly is well-being? Here are some common definitions:

A good or satisfactory condition of existence; a state characterized by health, happiness, and prosperity; welfare
The state of being comfortable, healthy or happy
Someone's well-being is his or her health and happiness

There may be no absolute consensus regarding a single definition of well-being, but most would agree that well-being should combine the presence of positive emotions

Work-related ill health

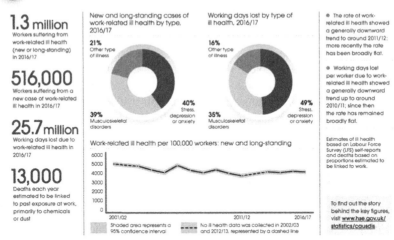

Work-related stress, depression or anxiety

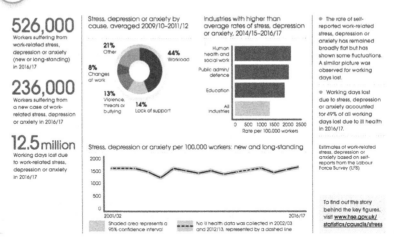

Figures for work-related ill health, particularly relating to stress, anxiety and depression, provide cause for concern as they show little sign of decline.

and feelings such as happiness or a feeling of good health with the absence of negative emotions such as anxiety or stress.

For now, I'm going to subdivide well-being into two categories, physical and emotional. There are, however, strong links between physical and emotional well-being and these remain somewhat underexplored in terms of workplace health and safety.

It's worth considering just how our state of well-being can impact health and safety, and inevitably, it's complex. Examples could include:

- Distraction due to worries about a relationship external to work or due to concerns about job security.
- Poor health can lead to long-term, chronic injuries or conditions, a shortened and less effective working life and a greater potential for error.
- Mental ill health could lead to short-cutting, difficulty in concentrating or an abandonment of procedures.
- It's possible that people with extremely high well-being scores could also be distracted due to their focus on highly satisfying out-of-workplace lifestyle factors.

As a general principle, it is considered significantly beneficial to have a workforce with high ratings for both physical and emotional well-being.

The big issue here is how, as an employer, is it possible to make a substantial impact on an individual's well-being, when many of the factors that influence a person's well-being are external to the workplace?

Before looking at that, I'd like to dig a little deeper into what factors influence a person's well-being. Maslow's hierarchy of needs provides a convenient way to examine some of the key and most influential factors.

At the foot of Maslow's pyramidal view are our physiological needs. These are the basic physical requirements for human survival, and in most cases, these needs are met within our society, though there is increasing evidence that some workers will not even attain these basic physiological needs. Should these requirements not be met, we cannot function properly, and our personal and working lives will be adversely affected.

Physiological needs are based on survival and include

Sufficient food and drink
Adequate sleep
Clothing and shelter
Reproduction

Safety needs form the second sector of Maslow's pyramid.
Safety needs include:

Personal security and safety
Financial security
Job security
Physical and mental health
Protection against variable and intermittent external factors

After physiological and safety needs are met, Maslow's third level of human needs relates to a requirement of social belonging.

Social belonging needs include:

Friendships
Family
Love
Belonging to social groups.

The next rung on Maslow's hierarchy moves to our need to be accepted, respected and valued by others and ultimately a feeling of self-esteem and self-respect. Key aspects include:

Desire to be accepted and valued by others
Desire for status and recognition, attention or prestige
Self-esteem
Self-confidence
Competence and assertiveness

Maslow's original pyramid was topped by self-actualization, or a realisation of full potential. In later life, he explored a further dimensional level and expressed more in terms of spirituality and a wider connection to nature, other species and mankind's place in space and time.

Maslow's hierarchy still provides an interesting basis for assessing well-being, but the modern world imparts a wider range of influences on the individual of both a positive and negative nature and these can have a profound influence on our well-being. Some of the original building blocks in his hierarchy have been distorted and require reformation and redefinition.

PHYSIOLOGICAL NEEDS

Our basic physiological needs haven't changed much in actuality, but they have changed significantly in terms of expectation. We still require food and drink, shelter, clothing and sleep, but over time, our expectations of what constitutes acceptable levels of these basic needs has changed significantly and the majority of society expects a much higher standard of these basic living essentials than at any other time. This, of course, reflects in our feeling of well-being. The bigger the disparity between what we perceive we require in terms of our basic physiological needs and what we actually have, the greater the likelihood that we will feel dissatisfied and unhappy. It's an expectation gap that has been driven by modern media and communications.

We may sometimes wish we could realign these expectations, but as managers, we need to at least be aware that this gap in expectations can be a contributory factor in low levels of well-being amongst a workforce.

Much evidence also points to lower-paid workers having increasing difficulty in meeting their basic physiological needs. This can partly be explained by reference to increased levels of expectation, but there is also recognition that some people on low incomes genuinely struggle. This can be evidenced by the increase in the numbers of people using food banks and the increase in so-called 'payday' lenders. A small but increasing number of people in work are homeless, with figures showing a 50% rise in the number of people sleeping rough in just 3 years.

SAFETY NEEDS

The immense pressures of modern society have also had a significant impact on our safety needs. Our personal safety is probably at a higher level than it ever has been, but the perception of many people is that the world is a very dangerous place, with higher levels of personal risk than ever before. The media, of course drives this. It's hard to watch TV these days without coming across programmes on policing, traffic cops, murderers and so on. The predominance of this type of programme, and the quick and wide-ranging publicity given to acts of crime that involve threats to our personal safety – including terrorist attacks – leads us to believe that our society, and our place in it, are under threat like never before. History shows us that there have been many times during which personal safety was at a much lower level, but we must be aware that the perception many people have of a dangerous world will affect their underlying feeling of security.

Our levels of financial and job security are now extremely variable, with lower-paid workers clearly under more pressure from these factors. Zero-hour and short-term contracts might be popular with employers, but they don't provide much job or financial security.

Likewise, in terms of our finances, there is currently a huge divide between rich and poor, with inequality in our society seen by most as a major issue. Those affected most currently appear to be workers on low to middle incomes. Changes to a whole raft of financial issues, from the student loans scheme to a wholesale overhaul of the benefits system, have probably impacted manual and semiprofessional workers more than any other group.

A very significant issue is our health – both physical and mental. I deal with this in more detail elsewhere, but as a society we currently face major problems with a range of health issues that are in many cases directly linked to societal changes that Maslow could never have predicted.

SOCIAL BELONGING

It's in this sector of Maslow's hierarchy that some of the most significant changes can be observed. The family has always been a backbone of stability for many people, yet we now have record numbers of single parents and broken families. For many people, this breakdown in relationships and family structure imparts a range of pressures external to the workplace that have potentially significant effects on their well-being and their subsequent ability to work effectively and safely.

However, perhaps the biggest influencer in this respect is the relentless advance in communication technology and our complex relationship with social media. Though both provide undoubted benefits, there is increasing professional concern about the detrimental effects of overuse of mobile technology and the social problems arising from the use and misuse of social media. This may be reflected in a work environment through overuse of mobile technology whilst at work and subsequent distraction or bullying through social media.

ACCEPTANCE AND SELF-ESTEEM

Our desire to be accepted and valued by others is a very strong driver for many people at work. This is closely linked with our feelings of self-esteem and self-confidence, which tend to be higher when influenced by the positive feelings of others towards us. This is very different from our desire for status, recognition, attention or prestige, which tend to be ego-driven and in many ways is exemplified by the plethora of reality TV shows offering so many people their so-called 15 minutes of fame. These factors can be distracting and remove focus from work objectives, alongside detracting from overall team goals at the expense of individual priorities. The link with safety concerns whether we are acting in self-interest or in the best interests of the team/ workplace as a whole. Confident, competent and assertive members of staff who are motivated by team success and the positive feelings of others are more likely to act in the best interests of the organisation than people who are motivated more by the egocentric desire for recognition or prestige. These are issues that can be addressed to some extent through training and team building.

REALISATION OF FULL POTENTIAL

You don't need to be CEO of a major multinational to feel as though you've reached your full potential – this can be achieved at any level.

In order to attain a high level of well-being, we need to consider the following:

Physical well-being
    Fitness
    Health
Emotional/mental well-being
    Happiness
    Stable personal, family and work relationships
    Positive emotional/mental health
Peripheral well-being
    Financial security
    Enjoying engaging activities and interests

There is inevitably a great deal of individual interpretation as to what merits a satisfactory level within each of these components. For example, one person's version of financial security might sit very uncomfortably with another person's view, but whether you feel financially secure with £100 in the bank or whether you need a portfolio of blue chip investments doesn't really matter – what's important is whether you have a feeling of financial security. It's in the interest of all organisations to find ways of helping their workforce get as close as possible to realising their full potential.

We also need to look very closely at the comparison between traditional ways of engendering a feeling of well-being, and what could be achieved now. For example, almost half of the respondents in a large 2017 survey stated that they would be willing to use an app if it helped in this respect, although about 40% of those people who said that they have taken time off or altered their role due to poor well-being would not want to share information with their employers. There's clearly a case for examining the use of mobile technology to promote well-being alongside more traditional methods.

Before considering what can be achieved at workplace level with regard to well-being, there are a number of major issues that require consideration:

- It's relatively straightforward for bigger organisations to make the sorts of provision I mention below, but recognisably more difficult for smaller employers.
- On the other hand, smaller employers can engage in a more personalised way with their staff, which may have advantages and disadvantages.
- Issues of trust require addressing alongside developing well-being programmes.
- Due consideration must be given to age, gender, ethnicity, culture and disability.
- What employers think will help to promote well-being might be substantially different to the viewpoint held by their employees, so active and thorough engagement is essential.
- Strategies and policies aimed at improving well-being should be developed as very long-term goals.

So, what can be achieved at the workplace level with respect to managing and improving the well-being of employees?

PHYSICAL WELL-BEING

HEALTH AND FITNESS

Physical fitness is a valuable and sometimes essential attribute for many jobs, but regardless of whether you're scaling scaffolding, lugging lagging or pushing a pen, if you're fit, you're likely to work more effectively, have more energy and work more safely, with bigger margins. In the construction process for London 2012, the fitness of employees was scrutinised more than it ever had been before. Workers had their fitness for work tested and programmes were set up to assist people who wanted or needed to be fitter for their work. In Cumbria in 2017, at least one construction firm was selecting employees based on their out-of-work activity and fitness level, the rationale behind this being that very fit workers are more productive and work more safely. Physical fitness is an essential ingredient of well-being.

Dave Birkett, a nationally recognised stonemason, said 'All the lads who work with me are involved in outdoor activities and they're all very fit. They work harder and they make fewer mistakes, their fitness gives them an edge and a healthy margin when performing more difficult tasks. In a lot of our work the days of the older guys who do nothing outside work and aren't very fit are gone'.

At a time when our workforce is ageing and a significant percentage of accidents involve workers over 60 (in 2016–2017, 25% of fatal accidents involved workers over the age of 60, despite this age group making up only 10% of employees), we need to focus on ways of ensuring our whole workforce is as fit as possible – it's easy to focus more on a younger generation when talking about fitness.

My own work experiences, plus the comments by Dave Birkett and the way in which the biggest construction projects are placing such high value on employee fitness, makes me buy wholeheartedly into the concept that fitter workers are safer workers. So how can we encourage and help our workforces to become fitter to complement and improve their feeling of well-being?

- As with so many initiatives, the example has to come from the top down, so management must fully support the concept and set an example. Creating a fitter workforce must be a stated and clear aim.
- Involve and engage the workforce; aim for joint ownership so the initiative is not seen as an imposition.
- Consider a clean and definite start point that could commence with, for example, fitness testing the entire workforce. Use these baseline results as a way of measuring progress. In the London 2012 construction project, fitness and health checks also revealed a range of pre-existing fitness and health problems. Once identified, workers were provided with appropriate help in order to deal with them, making this an even more valuable exercise.
- Provide appropriate professional advice for your staff. Whilst it's great to get people exercising more and becoming fitter, you've seen the health warning – 'consult a health professional before taking up new exercise ...' Hopefully a degree of common sense can prevail here, and those people who haven't exercised much in recent years or who have any underlying health worries will take some professional advice before commencing any new training programme.
- Provide support through membership programmes. Many larger organisations provide subsidised gym membership, but of course a gym is not the only way to keep fit. Subsidised memberships should be extended to a range of other sport and fitness areas such as indoor climbing walls and sports such as squash or badminton, swimming or yoga.
- Measure progress once a baseline is established, continue to promote fitness, engage with the workforce and celebrate success stories.
- Don't neglect the role that the great outdoors can play in terms of fitness. Walking is a great way of raising fitness levels and has the added benefit of fresh air and all the benefits that are acknowledged to come from the outdoor environment. There are many Walking for Health programmes throughout the country, and in cases where an organisation doesn't have the internal capacity to run a programme of walks, contacting The Ramblers or local conservation organisations will provide plenty of opportunities.
- Using the great outdoors for raising fitness levels doesn't stop with walking. All outdoor activities offer that great combination of exercise and environment. Consider providing structured opportunities for staff to try outdoor activities such as canoeing, rock climbing or orienteering. This has potentially beneficial team-building value, and those who enjoy the experience can move on to join local clubs.

- When promoting fitness, be aware that age, gender, disability and cultural background can all influence the type of participation you can expect, so take this into account in planning.
- There are strong links between feeling fit and healthy physically and positive mental health. There is strong additional evidence that attaining good physical health through using the outdoor environment has an even greater positive effect on our mental well-being.
- A 2017 survey showed that almost half of the people taking part would use an app that could help them with well-being. This provides an interesting reflection on the influence of mobile technology and a pointer towards strategy when considering different age groups and backgrounds.
- Strategies that aim to generate higher levels of fitness and physical health need to take into account gender, cultural and socioeconomic differences within any given workforce.

Physical activity is clearly linked with an improvement in physical and mental health through a number of factors:

Mood: Research has shown that physical activity has a positive impact on our mood. Studies showed that people felt more awake, calmer and more content after physical activity.

Reduction in stress: Being active on a regular basis has a beneficial impact on reducing stress. It can help manage stressful lifestyles and helps us make better decisions when under pressure. Studies on working adults shows that active people tend to have lower stress rates than those who are less active.

Improved self-esteem: Physical activity has a big impact of our self-esteem. Those with improved self-esteem can cope better with stress and demonstrate improved relationships with others.

Delaying cognitive decline: Sport and activity can help delay any decline in brain function. Studies show that those who are active are less likely to experience cognitive decline.

Helping with depression and anxiety: Exercise has been described as a 'wonder drug' in preventing and managing mental health. Many GPs now prescribe physical activity for depression, either on its own or in conjunction with other treatments. It is effective at both preventing onset of depression and in terms of managing symptoms.

Physical Health

I'm sure we'd all agree that good health is an invaluable, yet often taken for granted, asset that helps us enjoy and take full advantage of all aspects of our lives. We frequently don't realise just how much of an asset it is until something goes wrong. Our health covers way too many areas for me to focus on in any great detail, but suffice to say that it covers everything from colds and flu through to long-term, work-related chronic conditions, and from pulled muscles through to joint replacements.

If you don't benefit from good health, there is a chance that this will affect your ability to work safely. You can't possibly work as effectively when you've a heavy cold or flu. Degenerating knee or hip joints, and the pain and lack of mobility produced, limit your ability to react physically and lay you wide open to slips and falls. The pain or worry associated with ill health is a key distracter that will affect your ability to concentrate.

In many cases, particularly outside the biggest organisations and projects, physical health concerns often focus on short-term logistical issues such as eligibility for sick pay and getting people back to work quickly. It's also important to think in two more ways – interventionism and long term. Interventionist strategies aim to identify and treat health problems at the earliest stage possible, while long-term strategies look at the bigger picture and consider the long-term effects of health issues on both the employee and the employer. A classic example comes from one of Carlisle's leading food processing companies. They employed the services of a GP and physiotherapist on a regular basis to identify and treat health issues at an early stage. The company reported savings of three times the cost of providing the service, along with a reduction in the incidence of long-term sickness and payout costs. Academic research illustrates the same point – that the costs of early intervention and support are only about 30% or less of the overall amount saved, making sound economic sense.

- Fitness and health are clearly very closely linked, so it's worth considering a combined health and fitness approach.
- Once again, it's essential that management set a clear target and lead by example, making the link between ill health and its potential effects on health and safety.
- A health check for all employees provides a starting point from which improvements can be measured and can identify health issues that may not have been known about by either employer or employee. This can be performed at relatively low cost through individual practitioners.
- Wherever possible, look for early intervention strategies such as the provision of access to a GP and physio on a regular basis, which are clearly proven to save money in the longer term.
- Establish an appropriate culture and policy regarding contractible illness at work. Is it really acceptable to encourage workers to soldier on when they are suffering from a severe cold or flu? At best they will pass the illness to other workers, and at worst their ailment will make them unsafe.
- Take advantage of the government's Fit for Work programme where possible.

EMOTIONAL/MENTAL WELL-BEING

A 2017 study by Randstad, one of the world's leading recruitment agencies, revealed a range of important and disturbing trends regarding mental health within the construction industry.

Some 3,400 respondents contributed to the survey, which revealed that 34% of construction workers had experienced a mental health condition in the last 12 months, while 73% thought that their employers did not recognise early signs of mental health deterioration. For women, almost half of the respondents said that they were currently dealing with poor mental health, and the figures for loss of sleep and absence from work were both approximately 10% higher for women.

Twenty-nine percent suffered a loss of sleep due to stress at work, and 20% stated that they had increased their consumption of tobacco or alcohol as a result of stress-related conditions, adding to the evidence of a demonstrable link between poor mental health and subsequent deteriorating physical health.

The age at which the respondents experienced mental health problems revealed a significant trend, with 43% of 18–25 year-olds having experienced a mental health issue in the preceding 12 months compared to just 15% of those aged 66 and over.

The research produced some telling information regarding the relationship between employer, employee and finding solutions. Workers are reluctant in general terms to discuss or report mental health issues – almost half of those who took time off work due to mental health problems did not reveal the true reason to their employees. Only a third of those aged 35 and over would be willing to talk to their employer about taking time off due to mental health issues, whereas on a positive note, two-thirds of 18–25 year-olds said that they would be happy to do this. However, the mechanisms are often not in place, with well over half of respondents stating that the support structures were inadequate in their working environments.

By way of offering solutions, 43% said that they would prefer an anonymous helpline, 27% said that they would like to see a named person on site and 24% said that they would like in-work training.

The results of this and many other surveys demonstrate some simple facts:

* Mental health issues are growing within our workforces and are leading to millions of days being taken off work at a substantial personal and commercial cost.
* More employers are taking mental health issues seriously, but many – probably over half – are not able to recognise mental health problems and offer no structured support.
* The younger the workers, the more likely they are to suffer a mental health problem, while in some industries, women are significantly more likely to suffer from stress-related problems than men.
* Depending on age, between one-third and two-thirds of the workforce do not feel able to discuss mental health issues with their employers.

HAPPINESS

The John Lewis Partnership is well known for placing the happiness of its employees at the very top of its management aims. A happy, engaged workforce rewards its employer with loyalty, commitment, extra effort and a better engagement with customers. Put simply, a happy workforce creates a better and more profitable business, a statement backed up by modern academic research.

Recent survey statistics illustrate some interesting data regarding workplace happiness. Approximately 10% more managers are happier than nonmanagers, with a similar gap showing when they were asked if they felt trusted and empowered to make decisions by their employer and when asked if they felt pride in their work and are doing something worthwhile.

Perhaps more significantly, the least happy group were female workers and those between the ages of 19–24, regardless of level within a business. These groups were more likely to be underperforming and were not being used to their full potential.

For me, the most significant statistic was that the construction, transport/logistics and architecture/engineering sectors showed the lowest levels of happiness. Only 46% of workers in the construction industry said they were happy at work in comparison to a national all-sector average of 60%, and at the other end of the scale, the hospitality and nonprofit sectors showed scores of 74% and 75%, respectively.

When asked what measures are needed to increase happiness in the workplace, 32% asked for a better work/life balance, 24% suggested more flexibility in working hours and 12% suggested more workplace social events.

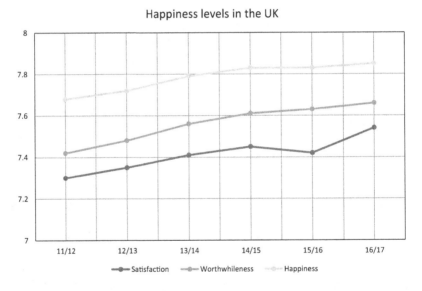

Happiness levels in the UK

The Office for National Statistics shows happiness levels steadily increasing, while HSE figures show contradictory, recently increasing rates of stress and related conditions.

So what's happiness got to do with health and safety?

Firstly, health and happiness are interlinked – especially with mental well-being. Happy people are less likely to take time off work due to mental health issues, and they are likely to engage more with colleagues and feel confident enough to express themselves with greater clarity.

Happy workers have fewer worries and negative thoughts than unhappy workers, and can be less prone to distraction and better able to concentrate on tasks.

Happiness and health and safety form a significant partnership in which increased levels of happiness can lead to a more effective, stable and safer workplace.

We can help create a happy workforce through the following means:

- Set a clear goal of creating a happy workforce. Make the goal clear and simple and engage with staff. You won't find many businesses whose goals include the creation of a happy workforce, so use this as a selling point to help your staff buy into the concept.
- Management must lead by example and create the right underlying culture and climate.
- Engage with the workforce to find out what makes them happy/unhappy at work, and take appropriate action.
- Provide opportunities for team building and social events, make them varied and inclusive and attract ideas from the workforce.
- Use performance management/reviews to check progress individually and use a monitor/review process with a clear score indicator.
- Set up a mechanism to allow staff to air concerns about things that are making them unhappy at work, and allow them to make suggestions as to how to make their working life happier.
- Provide mechanisms for the workforce to air out-of-work issues affecting their level of happiness. This could involve access to counselling, for example.

STABLE RELATIONSHIPS

It's extremely difficult to influence the stability of relationships that predominantly or wholly take place away from the workplace, yet our relationships outside work are critical to our well-being and to our ability to perform effectively at work. In my survey, one in four people said that their ability to concentrate at work had been adversely affected by personal relationship problems. It's clear to me from the survey, from other academic research and from personal experience that stable relationships outside work provide a solid platform that enables people to be both more effective and safer in their work environment. The survey also showed that although only just over 20% of respondents stated that personal relationship problems sometimes affected them at work, 55% said that they were aware of other people within their organisations who were affected by those sorts of problems.

Relationships within work environments also benefit from stability, and this is more within our control. We can use team building and social events to help build relationships and it's easier to be aware of where problems lie.

As managers, what can we do to help?

- Make it a clear management objective to provide help where possible to workers with personal relationship issues both outside and within the workplace.

- Develop a culture in which it is acceptable and normal to discuss problems with work colleagues where appropriate. Survey results suggest that very few people discuss relationship problems with work colleagues and this is a substantial barrier. Employing a professional counsellor for an hour to provide a lecture to the workforce on this issue would be a positive step.
- Provide dedicated support time through professional counsellors for staff. This could be a short visit from an appropriate person once a month or once every couple of weeks. This support should be provided just outside working hours, so that attention is not drawn to individuals wanting to take advantage of it. Evidence suggests that the majority of workers will not discuss personal problems at work and that the figure for male workers is higher than for female workers, so providing any route through which people can share these sorts of problem is valuable.
- The provision of counselling could be undertaken by a dedicated and trained member of staff; 27% of those surveyed in a construction industry survey said that they would like to be able to discuss mental health issues with a colleague trained and dedicated to do so.
- Provide access information to third party assistance, for example, Samaritans, Relate.
- Provide a stable and supportive work environment.
- Remember that a lack of stable relationships can form an important part of other mental health issues.
- Look for and get involved in schemes such as Mates in Mind.

POSITIVE AND STABLE EMOTIONAL/MENTAL HEALTH

Mental health issues may affect how we feel, think and behave and to a greater or lesser degree will limit our ability to cope with work, relationships and life in general. In the workplace, other employees may be affected as well, with additional workloads to pick up and the potential to impact an organisation's productivity and profitability through overtime costs and the recruitment of cover. Absence from work due to mental health issues is thought to cost the UK economy £26 billion per annum, but we can't be sure of its effect on health and safety other than to say that it is undoubtedly a major contributor to some of the key issues that are known to be the cause of many accidents, such as distraction and a lack of concentration and focus.

Virtually half of all long-term sickness is due to mental health problems and it's recognised that these impact more heavily a person's ability to work effectively and safely than any other illness. It's also recognised that mental well-being statistics demonstrate a higher average cost per individual than physical injury in terms of time off work, treatment and so on. Figures from the Mental Health Foundation suggest almost 15% of people are likely to experience mental health problems in the workplace, with women in full-time employment about twice as likely to experience common mental health problems as men in full-time employment. In total, just under 13% of sickness absence days are attributable to mental health conditions. MIND reports that approximately one in four people in the United Kingdom will experience a mental health problem each year. These figures appear to be seriously underestimating the scale of the problem within certain industries, such as construction.

For many years, mental health has been stigmatised to a great extent and it's only relatively recently that high-profile public figures in sport and other walks of life have talked more openly about how mental health issues have affected them, bringing mental health firmly into the public eye, but also starting to remove its attached stigma.

Mental health issues can appear a result of a combination of factors in both our working and private lives, as a result specifically of factors in either, or can simply happen. The most common issues are depression, anxiety and stress. There are many causes, including relationship or financial problems, workload or problems with colleagues, lack of job security or a fear of change. The scale of mental health problems is huge, and alongside the types of causes listed above, there is increasing evidence that the use of social media and information technology is fuelling a surge in mental health problems, particularly amongst young people. Many organisations are pushing for action to be taken to promote the positive effects of social media alongside mitigating the negative effects. The Royal Society for Public Health (RSPH) is working alongside the Young Health Movement, for example, on this topic, yet on the same web page you'll find the RSPH stating that Instagram and Snapchat are the most detrimental to young people's mental health and well-being, while just down the page there's a link to follow the Young Health Movement on Instagram!

The government set out its No Health without Mental Health strategy in 2011, one of its key aims being to get people to recognise mental health issues in the same way that physical or biological health is dealt with. In 2016, the Prime Minister pledged 'to tackle the stigma around mental health problems', but arguably failed to provide the funding to match the aim.

Some sectors have developed their own schemes to tackle workplace mental health issues, the Mates in Mind programme developed by the construction industry being a perfect example. This scheme is in response to the recognition that up to 70% of days lost from work in this sector could be due to mental health issues and that the industry had a level of corporate social responsibility to meet. This initiative came up with some other sobering statistics, such as the fact that the suicide rate amongst construction workers could be 10 times the rate for industry fatalities. The scheme promotes the concept that it's good and acceptable to talk to work friends and colleagues about mental health issues and equally importantly to be a good listener.

Although good work has been going on for some years now in terms of recognising and treating mental health disorders, the statistics show that rates of stress, depression and anxiety are broadly the same as they were in 2002. Mental health disorders are not industry specific; they are societal problems that require a much wider appraisal. Our task is first to limit the role that the workplace plays in creating stress, anxiety, depression or other mental illnesses and second to provide ways of helping our workforce to cope with mental disorders that arise from out of work problems or a combination of those and workplace issues. In order to do this, we should:

- Treat all health issues, including mental health, as you would safety issues. Make them a priority like health and safety, and establish this amongst your workforce. This requires a strong lead.
- Create an accepting culture that has a positive understanding of mental illness and that aims to support those affected.
- Try to tackle issues of mental health at an early stage by understanding and acting on the early signs of mental illness, and recognise how early intervention can make a significant reduction in costs for the employer, as well as helping the affected person to enjoy a normal working life sooner.
- Promote positive well-being, engage with external speakers and advisors.
- Deliver mental health awareness training to management teams to give them a greater understanding. Look for opportunities to roll this training out across the entire workforce.
- It's great to encourage people to talk about mental health issues more openly in the workplace, but it's equally important that people have good listening skills in response and training may be required.
- Tackle stress in your workplace – aim to create a stress-free environment; involve your workforce.
- Train a member of staff to act as a workplace counsellor and ensure everyone knows who the person is and how to contact him or her.
- Provide appropriate training so that everyone knows the signs and symptoms of stress, anxiety and depression.
- Carry out benchmark and follow-up surveys and monitor progress.
- Involve staff in matters that may stress them; find solutions together.
- Investigate and join schemes such as Mates in Mind and promote positive mental health.
- Be proactive about preventing stress and embed this in your decision-making processes.
- Provide the right support, including a positive return-to-work strategy that places the importance of a person's well-being ahead of the importance of returning to work.
- Provide easy-to-access information on relevant contact details for organisations such as MIND, Samaritans and so on.
- Look at the chapter 'A Connection with Nature and the Great Outdoors' for alternative ways of tackling mental health issues.

PERIPHERAL WELL-BEING

Financial Security

What constitutes financial security varies widely from individual to individual, but worries about personal finances are very real for many people and can cause considerable distress. This is yet another issue that is predominantly out of the workplace, and there's a limit to how much any employer can do to help. However some positive suggestions outside pension provision and salary would include:

- Offering financial management support through a third-party professional – this could take the form of an occasional free advice session with follow-ups to be arranged between the two parties as required.
- Offering counselling through a third-party professional on a similar basis.
- Providing support and guidance within the organisation if appropriate personnel are available. This is a possible opportunity for training someone to offer support.

ENJOY ENGAGING ACTIVITIES AND INTERESTS

Most leisure interests and activities take place outside the workplace, but traditionally, many larger employers have provided assistance in this respect through, for example, a range of sports and social clubs. Some workplaces have strong affiliations with activities or sports that the business has interests in, and there may also be scope for businesses to explore links with partners in terms of activity and interest provision.

In additional to the more traditional activities and interests, employers should also look at ways of establishing links with volunteer groups, for example, local conservation organisations. There's a lot to be gained from encouraging voluntary work and engendering a sense of environmental responsibility. This has a range of excellent side effects, including gaining a better understanding of sustainability and a demonstrable practical involvement that's great for image, too.

- Retain sports and social clubs where they exist and look to broaden interest groups.
- Explore working with partner organisations in terms of widening provision.
- Seek to work with local volunteer groups – many local conservation organisations, for example, would be delighted to have extra help and there are distinct advantages in helping create a greater awareness of sustainability and in the provision of goodwill and promotional opportunities.
- Be aware of the interests of different age groups, cultures and genders when looking to make any provision for activities and interests.

STATISTICS

Since 2011, the Office for National Statistics has been measuring personal well-being through its annual surveys in the United Kingdom. The results show a steady increase in levels of personal well-being, though this is somewhat skewed, as the measurements in England have driven the figure upwards, while levels of personal well-being in Scotland and Wales have remained static.

On the other hand, we have witnessed a slight increase in the levels of stress, depression or anxiety that resulted in time taken off work during the same period.

So, on the one hand, we are being told that levels of well-being are improving, but on the other hand, the incidences of stress, depression and anxiety – all closely linked with well-being – are actually increasing.

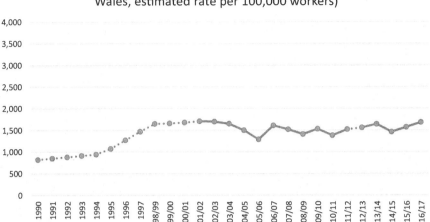

Rate of self-reported stress and related conditions (England & Wales, estimated rate per 100,000 workers)

The Office for National Statistics shows happiness levels steadily increasing, while HSE figures show contradictory, recently increasing rates of stress and related conditions.

THINGS TO WATCH OUT FOR

Looking ahead, there is clearly some significant momentum building regarding well-being, with more and more businesses being set up to offer advice, assessments and direct assistance and increasing numbers of employers becoming aware of how the well-being of their employees can affect their businesses. However, there are a number of concerns that I have.

The first is sustainability – how sustainable are the strategies we use to improve well-being going to be? Improving levels of well-being might be possible as a short-term, measurable goal, but the real benefits are in the long-term establishment of a set of conditions and cultures that help our workforces to gain, then retain, a good level of well-being. Succeeding in this will produce a more permanent reduction in accident, incident and near-miss figures.

The second is the relentless advance of the influence of mobile technology and social media. Increasingly, a range of professional figures of influence are becoming concerned about the effects of social media on society in general and in particular on young people. Concerns include direct effects such as sleep deprivation and addiction to mobile devices, cyberbullying, access to pornography and other unsuitable material and aggressive behaviour. There are also many indirect effects that have the potential to become longer-term problems such as lack of direct socialising and an inability to read nonverbal behaviour, difficulty forming cooperative and more widely inclusive relationships and long-term addiction to the internet. A former Facebook executive, Chamath Palihapitaya, expressed 'tremendous guilt' over his work on 'tools that are ripping apart the social fabric of how society works' and continues to say, 'This is a global problem. It is eroding the core foundations of how people behave by and between each other'.

From a health and safety perspective, there are a number of issues that raise concern, mostly concerning distraction and loss of concentration, though in the long term it's possible that behaviour changes such as difficulty working cooperatively and focus on individual rather than team could have an impact.

On the other hand, the use of social media and communications technology has, of course, many positive benefits and this needs to be factored into the ways in which we engage with employees in the future.

My third area for concern relates to demographics. There is likely to be an increase of approximately 8% in the coming 10 years in the numbers of workers in the 60+ age bracket. More people of this age and older are likely to stay in work longer due to the rise in the pensionable age and concerns about pension levels. This means a big increase in the number of workers who are most likely to be involved in accidents at work, and this will mean a specifically focussed effort will be required, not just in terms of well-being, but in terms of all the different factors described in this book.

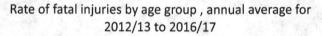

Rate of fatal injuries by age group , annual average for 2012/13 to 2016/17

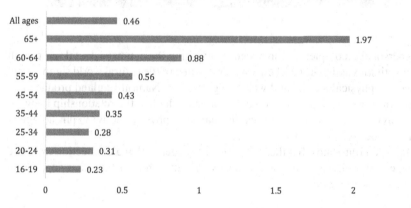

The rate of fatal injury exponentially increases with age, with workers aged 60–64 having a rate of almost double the all-ages rate, and workers of age 65+ having a rate of over four times the all-ages rate. While this age gradient in rate is most strongly seen in agriculture, it is also present across a range of other sectors too. With a predicted increase in worker age over the next 10 years, it's clear that employers must take account of and devise strategies to manage the anticipated increases in accidents and ill health.

Using the Great Outdoors to Improve Well-Being

It's long been known to outdoor education professionals that activity in the outdoor environment brings a wide range of positive well-being gains. These range from an uplifting feeling of euphoria when exposed to magnificent views or spectacular sights, through to benefits to one's fitness through activity.

In many ways, the evidence for the benefits of the great outdoors has been there for all to see for centuries – after all, where do most of us go for our holidays?

I'm personally completely convinced that visiting the countryside, taking part in outdoor activities and gaining a better understanding of the natural world significantly enhance our physical and mental well-being. In 2016, Natural England produced an evidence briefing entitled 'Connection to Nature', a look at the relationship between natural environments and mental health, learning, physical activity, physiological health and obesity.

Its research illustrates that there is emerging evidence that a connection to nature can make a significant difference to our well-being, though thorough supporting evidence is still required.

- Most existing studies show that spending time in or being active in natural environments is associated with positive outcomes for attention, anger, fatigue and sadness, higher levels of positive affect and lower levels of negative affect (mood/emotion) and physiological stress.
- Positive evidence relates to the impacts of activities in natural environments on children's mental health and their cognitive, emotional and behavioural functioning.
- One study found that regular use of natural environments has been shown to be associated with lower risk of poor mental health.
- A study of the behaviour of children with Attention Deficit Hyperactivity Disorder in different environments found better concentration in woodlands in comparison to urban places.
- Most studies find that greater degrees of natural environment around the home have a protective effect on self-reported mental health and are also associated with reduced risk of stress, psychological distress, depressive symptoms, clinical anxiety, depression and mood disorders in adults.

- Data analysis from over 10,000 people in England suggests that people report lower mental distress and higher well-being when living in urban areas with more greenspace in comparison to when they lived in urban areas with less greenspace.
- A study carried out in deprived populations in urban Scotland found more greenspace was associated with lower levels of self-reported perceived stress and improved physiological stress.
- In children, a greater quantity or proximity of natural spaces around the home or school is significantly related to improved cognitive performance and reduced incidence of behavioural issues.
- Living environments with a greater amount of greenspace are associated with reduced likelihood of depression and anxiety amongst older people.

Please see the chapter titled 'A Connection with Nature and the Great Outdoors' for more details.

17 A Personal Perspective
Part 3

In 1990, I set up a mountaineering school, providing guiding on mountain walks, scrambles, rock climbs and snow and ice climbs, combined with skills training courses. Every working day presented a range of potentially serious hazards, many of which could affect myself and the other guides I worked with far more than the people we took out.

Hazards such as extreme cold; heavy rainfall; severe winds and rough, loose and slippery terrain along with swollen streams and limited visibility due to hill fog were often present. Just occasionally, hot weather presented a range of hazards, but these days were sadly few and far between! When climbing, always going first (leading) exposed us to the risk of falling from height; stonefall was an ever-present worry, as was the onset of bad weather. In winter, there were the additional problems presented by avalanche danger; white-out conditions and steep, icy slopes.

Such a wide range of regularly presented hazards meant that guiding in the mountains was potentially a highly dangerous job. The hazards mountain guides are exposed to are the same as those that most amateur climbers are exposed to, except as mountain guides you are always exposed to the higher risks of going first and are exposed to the objective dangers found in the mountains for much longer periods, necessitating a keen eye for danger, constant observation and decision making, and a deep understanding of the mountain environment. A periodic slice of luck doesn't go amiss either!

My career in the mountains wasn't without incident, though, and it eventually led to years of discomfort and pain.

The writing was on the wall over 20 years ago, when I was lying in a hospital bed following keyhole surgery on my right knee to smooth out a cartilage tear. The registrar who came to see me following the operation was a delightful young chap, but his easy demeanour hid a message with the dagger point of a stiletto. 'I'd recommend that you consider your options in terms of the work you do, Mr White'. That slightly hazy period when you're recovering from a general anaesthetic isn't a great time to be told anything, but I recall with great clarity the hammer blow I felt as his words sank in. Climbing and the outdoor world provided both my work and my fun, and, outside family, it was everything. Having to stop working as a mountain guide seemed unthinkable.

So, I did what any pig-headed, macho outdoor type would do and vowed to prove him wrong. What does he know anyway? He's seen a couple of x-rays and thinks he can map out the rest of my life based on that! I knew better, of course, and was supremely confident that I could carry on as before. A few days later I was in Cornwall, walking along the cliff-tops, doing a little gentle climbing and wondering what on earth the fuss had all been about. If only I'd listened to him.

The ensuing years saw two more knee arthroscopies and a ripped quadriceps tendon as a result of a skateboarding accident (the less said, the better!) Descending hills with a heavy backpack became tortuous, the knee pain sometimes excruciating, and only regular painkillers and anti-inflammatories kept me working. I then started to develop some pain in my groin alongside knee pain, and went to see my GP about it. The young lady doctor suggested I could have a hip problem, and the ensuing x-ray suggested wear and tear in the hip joint. I was having none of that. What does she know, anyway? I carried on taking painkillers, ignoring the symptoms, confident that in some way, everything would be fine. After all, I was indestructible.

A few months later, I started to struggle. When playing sport, I had severe pain on the front of my thigh and groin, and at the end of longer walks, I could barely move. Looking back, I struggle to comprehend how I put up with the pain for so long and how I just carried on suffering without taking action; I now understand that this is due to a process of normalisation, where even severe discomfort can become the expected norm. I also find it difficult to believe how ignorant and short-sighted I was.

As the pain increased and my mobility decreased, I arranged another GP visit, but this time not for me to dismiss expert opinion again, but to find a surgeon who specialised in performing hip operations on people who were younger and active, and who wanted outcomes that maximised the opportunity of regaining an active life. I made an appointment to see a surgeon at Wrightington Hospital – the national centre for hip replacements. As I sat waiting to see the surgeon, I noticed how so many people in the waiting room walked like drunken sailors; the rocking motion that hip joint deterioration causes is uniquely identifiable. I then realised that was how I had started to walk, too.

The door creaked open. 'Mr White?' I strode positively into the office, trying hard not to look like the aforementioned drunken sailor. The surgeon looked at me impassively, examined the x-rays, glanced back at me and said 'Mr White, your hips are shot'. I felt like I had been. The x-rays showed severe wear and tear damage to both hips, and the only way forwards was a double total hip replacement. In the ensuing five months while I waited for the first operation, I deteriorated further, and walking half a mile became a real battle. I thought back to the young registrar who had advised me to change career 15 years before and wondered how I could have been so arrogant to think that I knew better.

The lessons to learn at this stage from my experience are pretty obvious, and need to be heeded by anyone with concerns about their health and well-being.

- Listen to expert opinion – you don't know best, they do.
- Take early action; don't just leave it until the pain and discomfort become normalized.
- Listen to the warning signs and don't pretend they're something else.
- Don't get by on painkillers; if you're having to do that, there's something wrong.

I had the two hip replacement operations 10 weeks apart, which was tough – just when you felt like you were making progress from the first one, it was back to hospital to have the same experience all over again. My bullish determination to get back to mountaineering saw me adopting a rigorous approach to rehabilitation and training. Seven weeks after the first operation, I went to Finland and walked kilometres through the snow in −20°C, got back onto skis on easy runs and drove a snowmobile 60 km to the Russian border. The second operation went more smoothly and comfortably than the first, and for once, I had learned some useful lessons that helped in the recovery process. Four months after the first operation, I did my first mountain walk and climb, and although I stopped guiding climbs, having come to terms with that limitation, I carried on guiding people on mountain walks and felt good in myself again.

About 6 months after the operations, I started to get some periodic pain in my right thigh. The hospital couldn't work out what it was, and its sporadic nature didn't point to any one thing in particular. During the first couple of years after the operations, I'd experience the thigh pain every so often, but it would always settle down and I'd enjoy a pain-free period. In the third and fourth year post-op, I started to feel very acute, localised pain in my thigh that sometimes seemed to be brought on by doing physical work in awkward positions, yet at other times came and went with no obvious cause.

I went to see a different consultant orthopaedic surgeon, recommended to me by a surgeon friend. He studied my x-rays for some time, and proceeded to explain them to me in the sort of detail that nobody had been prepared to do before. Both of the sockets that receive the hip joint appeared to be perfect. My left hip joint had a different design to my right – the implant was bigger, and it had a collar on it where it entered the bone. The right implant was thinner and shorter, and was set at an imperfect, slanting angle rather then leading vertically into the femur. My right leg, he assured me, was a couple of centimetres shorter than my left. I was advised that the pain could be resulting from the fact that one leg was shorter than the other, or that the right implant could be slightly loose, or possibly that its smaller size and bent aspect could be placing stress on the bone and therefore causing pain. I had a nuclear dye test, which seemed to confirm that it wasn't loose. I waited several months for the local podiatry department to spend a couple of minutes providing me with an off-the-shelf heel insert. It made no difference.

In the summer, 5 years after the original operation, I spent a week of my holiday barely able to walk, and I went to see my GP, who arranged for an x-ray and subsequently an MRI scan. Time elapsed and I was getting desperate. I went to see a very good physio who specialised in sports injuries. She worked on the pain site a couple of times, and I had a few weeks of virtually no pain, which was a revelation. I managed to climb Troutdale Pinnacle, a wonderful 100-m rock climb in Borrowdale, and felt like I had a new lease on life. The physio thought the pain might be resulting from nerve damage caused during the healing process, but it came back again. I had an MRI scan and to my surprise, my GP had asked for it to be done on my lower back, not my thigh. I was mystified.

Time had slipped by to early spring, and I was in considerable pain most of the time. I went to see another physio, who agreed that it was most likely pain from a trapped or damaged nerve. I told him I was going to Austria the following week to see my eldest son, who had spent the winter skiing. 'If I feel I can, should I ski?' came my question. The answer came back a resounding 'Yes – you're not going to do any more damage'.

I went to Austria and hobbled off the plane with a crutch, much to my son's chagrin, as he wanted a skiing partner. The first day, I watched him ski, and the second day, I somehow managed to get a pair of ski boots on and went up and down a beginner's lift a few times. On the third day, I downed plenty of ibuprofen and paracetamol and headed skywards in the amazing cable car that accesses the Dachstein glacier. It was magnificent – a windless morning with the jagged limestone peaks of the high Dachstein etched against a perfect azure sky. I managed about 10 runs before I'd had enough, and the thrill of skiing eased my pain as much as any painkiller.

Back down in the valley on the way to the airport, my phone rang. It was my GP calling to give me the results of my MRI scan. I listened expectantly, but was left open-mouthed and disbelieving when he told me that my thigh pain was the result of wear and tear in my back. I listened politely, but it made no sense at all and I didn't believe him.

Before I left for Austria, the original physio I had visited recommended that I should go back and see the surgeon who had originally performed the operation, and I flew back on Sunday with my appointment booked for Monday. Maybe he would be able to get to the bottom of this nerve problem – I certainly hoped so, as I had truly had enough of being in constant pain.

He asked me about my experience with my hip replacement, and I went through it all yet again, finishing with, 'I've been to see two different physios and they think it's a trapped nerve, but my doctor thinks it's my back'. The consultant remained noncommittal, but asked me if my right leg felt shorter than my left, and I had to say it did. 'I think I know what's wrong, but you'll need to get x-rays, I'll see you as soon as you're back'.

Back in the consultant's office once more, I sat down and waited eagerly for the computer to load the x-rays, but nothing could have prepared me for what I was about to see. The titanium implant that went down into my thigh had snapped in half about halfway down, overlapping, and leaving the upper portion sticking out at an angle into my thigh bone. It didn't take a consultant to see what the problem was! He intimated that it probably broke because it was too small for my body size and activity level, and being set at an angle meant it was subject to abnormal loading.

I was told I needed an emergency operation and in the meantime to use two crutches and take weight off my right leg, or there was a possibility that the fractured implant would in turn break my femur. The emergency operation was 5 weeks ahead, so I steeled myself for 5 weeks on crutches, but underneath the bravado, I was desperately worried about the outcome. Still, I'll never forget the consultant's face when I told him I'd been skiing the day before.

Five weeks later, I had a 4-hour operation to remove the old implant and insert a new one. This involved cutting a 180° section into the femur (effectively breaking my leg) in order to remove the old implant and drilling the femur out to receive a bigger and longer implant. The recovery process was much slower than for a normal hip replacement, but I was honoured when the district nurse announced, while removing the stitches, that it was the longest scar she'd ever seen. Ten months later, I'm sat writing this pain free, and though it's not perfect, I'm walking again and able to enjoy a reasonable amount of exercise knowing that although I might be a bit stiff, it's not going to hurt like it did.

Now that was quite a lengthy story, but it has a serious, undeniable and essential set of messages.

- Don't wait. Don't be a stubborn, I-know-better-than-you idiot like me. Seek help at the earliest opportunity.
- If something's not right, don't allow normalisation to occur. I'd got myself into a state of mind that viewed pain as acceptable, because I'd been in pain for so long. The normalisation process that subtly changes our perceptions over time took hold.
- Don't assume medical professional opinion is always right. It frequently is, of course, but if something just doesn't feel right, just doesn't make sense, question it. The physios acted in good faith when they told me the pain could be down to nerve damage. The GP acted in my best interests when he told me my back was the source of the pain, but they were both wrong.
- Get a second opinion. A surgeon friend told me to always get second opinions on medical matters. 'You always get two or three quotes for work on your house, so why not get two or three opinions on an ailment?' If you're not comfortable with your diagnosis, get a second opinion.
- I was self-employed during most of this period, but had I been employed in a conventional sense, the short- and long-term costs to my employer could have been significant. As an employer, it's in your interest to actively encourage employees to seek medical help at as early a stage as possible.

'Act now – don't wait' – that's the message to stress to your workforce.

18 A Connection with Nature and the Great Outdoors

The outdoor environment has long been linked with a feeling of well-being. We've been taking holidays in the countryside, along the coast and amongst the mountains for hundreds of years, enjoying the fresh air, the tranquillity and the views.

The natural world has the capacity to inspire and move us, and exposure to the natural world has been clearly linked with improved levels of well-being.

Outdoor professionals and educators have also recognized for a long time the very special role that the great outdoors has to play in terms of physical and mental well-being. This has recently been backed up by a series of reports prepared by Natural England on the link between natural environments and a range of outcomes, including physical activity, physiological health, obesity, learning and mental health. The reports found emerging evidence of multiple benefits, though there was often a lack of up-to-date, rigorous academic research. However, as I mentioned earlier, professional outdoor educators have long recognized the many and varied benefits provided by the outdoor environment, and I have no doubts that such benefits are extensive and, at a population level, extremely important. Some of the key findings in the reports included:

- At a population level, higher levels of exposure to natural environments are associated with lower all-cause mortality, rates of diabetes type 2 and cardiovascular and respiratory disease, and more positive maternal and pregnancy outcomes.
- A Scottish study showed that physical activity in natural environments is associated with a reduction in the risk of poor mental health to a greater extent than physical activity in other environments and that those who regularly used woods and forests for physical activity were significantly less likely to experience poor mental health compared with those who did not use such places.
- Compared with indoor activities, physical activity in natural environments is associated with greater feelings of revitalization and positive engagement; decreases in tension, confusion, anger and depression; and increased energy.
- Several studies suggest that people enjoy physical activities more when undertaken in greener environments. A systematic review found evidence that people were more satisfied following physical activities in the outdoors (compared to indoors) and reported a greater intention to repeat the activity at a later date.
- A review of older people's physical activity found that opportunities to spend time in natural environments were one of the factors which encourage participation.
- Green exercise programmes such as outdoor walking groups have been shown to increase activity rates and result in improved self reported self-esteem and mood states and are increasingly commissioned by health care providers. Evaluation of the Sport England–led 'Active England' Woodland projects found increases in engagement by previously disengaged groups.
- The estimated values of a proposed expansion of the Walking for Health programme were found to be: 2817 Quality Adjusted Life Years (QALY) delivered at a cost of £4008.98 per QALY. This was estimated to be a potential saving to the health service of £81,167,864 (based on life-cost averted) at a cost-benefit ratio of 1:7.
- Social return on investment assessments undertaken by Greenspace Scotland found a range of favourable cost-benefit ratios of health-related natural environment interventions, including the generation of around £5 of benefit for every £1 invested in a single health walk.
- A small but robust body of evidence suggests that natural environments provide exposure to the microbial diversity necessary for immunoregulation. It is thought that exposure to microbial diversity (including that from the natural environment) affects the human microbiome, which is linked to many health states.
- Most studies show spending time in or being active in natural environments is associated with positive outcomes for attention, anger, fatigue and sadness, higher levels of positive affect and lower levels of negative affect (mood/

emotion) and physiological stress. There is generally positive evidence relating to the impacts of activities in natural environments on children's mental health and their cognitive, emotional and behavioural functioning.

- A study found that regular use of natural environments has been shown to be associated with lower risk of poor mental health.
- Analysis of data from over 10,000 people in England suggests that people report lower mental distress and higher well-being when living in urban areas with more greenspace in comparison to when they lived in urban areas with less greenspace. Further analysis showed that moving to greener urban areas was associated with sustained mental health improvements.
- A study carried out in deprived populations in urban Scotland found that more green space was associated with lower levels of self-reported perceived stress and improved physiological stress.
- In children, a greater quantity or proximity of natural spaces around the home or school is significantly related to improved cognitive performance and reduced incidence of behavioural issues.
- Living environments with a greater amount of greenspace are associated with reduced likelihood of depression and anxiety amongst older people.
- Reviews of 'nature-assisted therapies' and 'green care' have found some evidence to suggest the activities may positively affect outcomes such as mood state, depression, dementia-related symptoms, frequency of negative thoughts and psychosis. Longer programmes appear to result in better outcomes.
- A Cochrane review of the benefits of conservation activities such as the TCVs Green Gym showed that exposure to natural environments, achievement, enjoyment and social contact were important pathways to positive mental health outcomes.

These reports add much weight to the often anecdotal benefits that outdoor professionals refer to.

The outdoor environment provides us with a number of distinct opportunities and benefits, as follows.

A FEELING OF CONNECTION WITH NATURE

It's quite hard to determine exactly what this means, and indeed it may mean different things to different individuals. For some people, it's a connection with our rural past; after all, it's not that long ago that a significant proportion of the population lived and worked in the countryside. Many of our basic instincts derive from ancestors who had to survive in an outdoor environment that offered both benefits and dangers, and we still feel those instincts today to some extent. The Biophilia hypothesis proposes that as humans, we have an innate tendency to seek connections with nature and other forms of life. Nature provides the basic rhythm of our lives: the seasons; warmth and cold; starlight and storm; food and the elements fire, water and earth. Many of us form some sort of virtual reality connection with nature through TV programmes, magazines and amazing photos, but to connect with nature by being outdoors and experiencing it for real provides a deeper and more meaningful experience altogether.

Close encounters with wildlife provide memorable and deeply moving moments.

FRESH AIR AND A HEALTHY ENVIRONMENT

The countryside normally provides a less polluted environment, with lower emissions levels from vehicles and industry, and a feeling of freshness that is regularly commented on. Within the outdoor environment, there are significant differences in this respect. That feeling of fresh air and a clean, healthy environment varies with location. In the United Kingdom, there's a tangible difference between, for example, the countryside of the Home Counties, the Lake District fells and the remote parts of Northern Scotland. Travel further afield to Greenland or the Alps and you'll experience air clarity of a different level altogether. In most coastal areas, you'll experience similar feelings, while research has shown that sea air is enriched by tiny water droplets that are enriched with salt, iodine, magnesium and trace elements that stimulate immune reactions by the skin and respiratory organs. The fresh air of the deep countryside, coast and high mountains simply makes us feel better, and that's a great thing.

WIDE RANGE OF ACCESS AND RECREATIONAL OPPORTUNITIES

The great outdoors provides us with a range of wonderful recreational opportunities, helped in the United Kingdom by the wide availability of access areas such as national parks, country parks, Forestry Commission access areas, common land, nature reserves and of course our unique and precious rights-of-way network. Wherever you are in the United Kingdom, and perhaps in contradiction to some views, you're always close to access to some sort of greenspace.

How land is used in the UK

Green Urban, 2.50%

Built On, 5.90%

Natural, 34.90%

Farmland 56.70%

The United Kingdom provides many opportunities for outdoor recreation and, as can be seen from the map, vast areas of open space.

UPLIFTING VIEWS AND CALMING LANDSCAPES

The great outdoors has the capacity to inspire and calm in equal measures. From gentle, rolling woodland whose green canopy exudes tranquillity and calm, to savage mountain landscapes that inspire a sense of awe, the United Kingdom has it all. Few of us will not have experienced these feelings at some time in our lives; in fact, research shows that the sense of wonder created by the natural world provides us with some of our most powerful memories. We also have an estimated 19,000 miles of coastline. National parks cover almost 10% of England and almost 20% of Wales. Woodland is now estimated to cover almost 13% of the United Kingdom. How much of the United Kingdom is built on? The UK National Ecosystem Assessment (NEA) analysed vast quantities of data and calculated that '6.8% of the UK's land area is now classified as urban' (a definition that includes rural development and roads).

The urban landscape accounts for 10.6% of England, 1.9% of Scotland, 3.6% of Northern Ireland and 4.1% of Wales, and 80% of the population lives in that area.

Therefore, almost 93% of the UK is not urban and contains just 20% of the population. But even that isn't the end of the story because urban is not the same as built on. In urban England, for example, the researchers found that just over half the land (54%) in our towns and cities is greenspace of some sort – parks, allotments, playing fields and so on. Furthermore, domestic gardens account for another 18% of urban land use and rivers, canals, lakes and reservoirs an additional 6.6%. Their conclusion was that in England, '78.6% of urban area is designated as natural rather than built'. Since urban only covers a tenth of the country, this means that the proportion of England's landscape that is actually built on is a tiny 2.27%. Even taking into account the sprawling nature of some development and the inclusion in these figures of greenspace to which there is no access, it's clear that the United Kingdom still has a huge area of countryside along with ample opportunities to explore it.

A CHALLENGING ENVIRONMENT

Some parts of the United Kingdom provide extremely challenging environments. The wilder parts of the Scottish Highlands and Islands and the Northern Pennines spring to mind, though there are many other areas that provide ample challenges without being truly wild. The significance of this from a recreational and developmental point of view is simply that these places exist. They exist as challenging areas for people to aspire to, so there's some sort of progression in terms of outdoor opportunities. These challenging environments are also home to a great diversity of wildlife and once again prove to be both inspirational and aspirational. Our challenging environments have also played the role of training grounds for some of the greatest explorers and adventurers – the successful first ascentionists of Everest trained in North Wales, and the outcrops of rock close to urban centres such as Sheffield, Leeds and Manchester have inspired some of the world's finest rock climbers.

A CHANCE TO OBSERVE AND LEARN ABOUT WILDLIFE

Whether you live in central London or on an isolated Scottish Island, the United Kingdom has a wide diversity of wildlife and an equally diverse range of conservation organisations. There are thousands of nature reserves, ranging from wilderness reserves in the Highlands to the 20+ nature reserves in London. Brockholes nature reserve is adjacent to one of the busiest M6 junctions at Preston. There is something for everybody, with hundreds of conservation organisations offering a wide range of events, walks and talks countrywide. A close encounter with our native wildlife is incredibly rewarding and it's worth enlisting the help of a local specialist to help get the most out of any visit.

SOCIAL AND COMMUNITY BENEFITS

Along with an increase in physical activity, outdoor recreation offers the chance to socialise, an important benefit in itself. For instance, bird watching incorporates several activities, including walking, interpreting visual and auditory input and speaking to other bird watchers. In general terms, the outdoor world is friendly and welcoming and often invokes a sense of camaraderie and togetherness. Outdoor physical activity such as walking can also increase pride in the community and its assets, as well as offering the opportunity to meet people with similar interests.

AN OPPORTUNITY TO GET CLOSER TO CONCEPTS SUCH AS SUSTAINABILITY

A sustainable approach to doing business is already a worthwhile and core principle adopted by many organisations. To appreciate the concept of sustainability, I believe it is very important to have a clear understanding of the natural world and the way that we fit into it. Understanding our development and history and having an appreciation of the effects that we are having upon the natural environment first hand can give your workforce a much clearer picture of why sustainability is so important.

OUTDOOR ACTIVITIES

The United Kingdom is a real hotbed for outdoor activities, with provision for everything from rock climbing to geocaching and from wilderness camping through to white water canoeing. The United Kingdom has a strong history of producing exceptionally talented outdoor sports people, and the opportunities to get involved are endless. Outdoor activities don't have to be risky and tough. Geocaching (using a GPS to find hidden caches) or Canadian canoeing, for example, offer more gentle activity in the countryside. Studies have shown that outdoor activities can lead to an increased confidence, improved creativity and better self-esteem. Natural settings rejuvenate and calm the mind, improve outlook and increase positive feelings. In contrast, artificial environments may cause feelings of exhaustion, irritability, inattentiveness and impulsivity, according to Resources for the Future. Outdoor time can even help you focus; 2009 research in the *Journal of Attention Disorders* shows that 20-minute walks through natural settings lead to improved concentration.

The key benefits provided by the outdoor environment include:

- Reduction in stress through a calm and peaceful natural environment
- Provision of space to think and clear your mind
- A feeling of good health promoted through fresh air and environment
- Improved concentration levels and help focussing on problems
- Improved confidence
- Improved fitness through exercise
- Greater awareness of sustainability issues
- A fantastic environment for the provision of team-building events
- Clear benefit of improved well-being

Key points for managers to be aware of include:

- The outdoor environment has the undisputed and well-researched potential to affect our mental and physical well-being in a positive way.

- Not every individual will benefit as much or in the same way from participation in activity in the outdoor environment. Some people will relish the challenge presented by an outdoor activity such as climbing, and others will not, but may enjoy walking or wildlife watching.
- Longer-term programmes are more effective than short-term ones, so avoid one-off events.
- Outdoor events and participation don't need to take a long time per event in terms of beneficial returns – much can be achieved in a short time. A one-hour walk is sufficient to carry benefits.
- There are many opportunities for creating goodwill alongside positive promotional opportunities.
- Financial benefits are well recorded. In Scotland, research in Glasgow demonstrated that for every £1 spent on the Health Walks programme, £8 of benefits were generated for society. Research carried out in Stirling and the Borders showed social return on investment ratios of £1:£9 and £1:£8, respectively. Although such benefit ratios may not be as high with regard to savings in the workplace, a projected £1:£3 or £1:£4 ratio nonetheless makes excellent economic sense.

The natural world clearly provides many benefits, but how do they translate into working towards a reduction in the incidence of accidents, incidents and near misses?

The key significant benefit is in terms of well-being. Evidence clearly demonstrates that our mental health, concentration levels and focus can be improved by using the outdoor environment in different ways. We can also look specifically to lower levels of stress and tension, revitalization and increased energy. There is no absolute and fixed correlation between any of these factors and health and safety. What we can say, however, is that if we improve the mental and physical well-being of our workforce, improve concentration levels and focus, reduce stress and create a happier, healthier group of people, we are directly tackling some of the key contributors to human error and helping our workforce towards better long-term health and fitness and a reduction in time taken off work due to stress and other mental health issues.

Finally, here are some practical ideas that will help you to involve your workforce in the great outdoors:

- As with any other strategy, this has to come from the heart and requires strong management backing, enthusiasm and leading by example. Don't forget that there are many additional benefits to your organization such as an increased awareness and practical understanding of sustainability and great promotional opportunities.
- To connect with nature, a good starting point is to contact your local wildlife trust. It will have opportunities for visits to nature reserves and for helping out on conservation projects. Doing something on a local basis is easier, and provides partnership, sponsorship and promotional opportunities. If your worksite is in or near a national park, you might want to contact their ranger service, and local National Trust offices may also be a good starting point. There are many examples of excellent, mutually beneficial working

relationships that have developed between businesses and conservation organisations.

- There are strong team-building opportunities here as well, and these don't have to involve days off work or weekends away. Hire a specialist walk leader to do something really interesting with your group – I know of examples where a group has hired a leader to undertake night walks looking for glow worms, or for stargazing when one of the meteor showers is in progress. Some fresh air, a walk looking at something totally different followed by a drink in the local pub makes an inexpensive and highly effective team event.
- Get involved in the Health Walks programme, a national scheme that encourages, promotes and provides opportunities for people to enjoy walking and make health gains as a result. A recent report by The Ramblers (provider of walks) and MacMillan Cancer Support provides evidence of the success of the scheme, evidence that can be rolled out to back up other forms of activity in the great outdoors.
- A report states that physical inactivity now rivals smoking as one of the nation's biggest health problems, currently being held responsible for 17% of early deaths in the United Kingdom and an estimated 6% of early deaths worldwide, making it one of the top four global killers.

Physical inactivity is also the principal cause of many common health conditions, including:

10% of heart disease cases
13% of type 2 diabetes cases
18% of colon cancer cases
17% of breast cancer cases

Research also reveals that being inactive can increase your chances of developing cancer, heart disease or having a stroke by 25%–30% and can take 3 to 5 years off your life expectancy.

There's a monetary price to pay as well, with estimates of a cost of £10 billion per annum through sick leave, health care costs and early deaths.

Adults should do at least 150 minutes of moderate physical activity such as walking every week according to the UK Chief Medical Officers, but some studies have suggested that only 6% of men and 4% of women actually achieve this. A third of people don't even manage 30 minutes a week.

The Ramblers report goes on to stress how walking can help you maintain a healthy weight, reduce blood pressure, increase fitness, improve balance and reduce the risk of falls.

The report also states that keeping active helps us feel both healthy and happy. People who stay active are less stressed, sleep better, have a 30% lower risk of getting depressed and have improved memories and sharper minds.

Walking is a perfect activity to promote through the workplace. The opportunities are virtually endless, it's already one of the world's most popular recreational activities and it's ideal for people who don't enjoy, or are reluctant to pay for, gym membership, for example. It's also ideal for individuals starting out with relatively poor health, as it's easy to build up from very simple beginnings.

A look at the Health Walks website will quickly point you in the right direction in terms of finding local walks, and you could also consider hiring a walk leader for a specific group. Walking – so simple, yet it works.

- You could consider running/arranging taster sessions in a range of outdoor activities. What you choose depends on what's available locally and your workforce, but there's a huge range. Most people can undertake activities such as rock climbing or Canadian canoeing, and they provide opportunities for teamwork and personal challenges alongside the team-building and well-being elements. At the opposite end of the spectrum, you could try orienteering or geocaching – both offer opportunities for learning a new skill (navigation/map reading) alongside healthy exercise in the outdoors. Contacting the national governing bodies of any particular sport is a quick way of finding appropriate providers. Running taster sessions in outdoor activities provides an immediate short-term gain in terms of team building, and longer-term benefits accrue from future participation or if the activity kick-starts a desire to get fitter.

Use the great outdoors to improve the overall well-being of your workforce and you'll reap the collective benefits for years to come.

19 Team and Relationship Building

Forty-four percent of people who responded to my health and safety survey stated that team building and social events would help to reduce stress and improve well-being in the workplace.

Team-building events are popular and varied both in nature and in efficacy. Events range from the more extreme 'walking on hot coals' type through to trying outdoor activities, to problem solving and 'Crystal Maze' type challenges and heavily learning-oriented events.

Using the outdoor environment as a vehicle for training managers and building teams took off in a big way in the early 1980s in areas such as the Lake District. In Scotland, John Ridgeway had his own version in the NW Highlands and companies sprang to life throughout the United Kingdom as businesses bought into the idea that running outdoor events could provide them with better managers and more cohesive teams.

However, many events had no clear aims and did not cater to the wide range of aptitudes and fitness levels present in an average group. Physical challenges and hardship might work for a small group of people as a motivator and in terms of self-development, but for many people, they do not. This was epitomised by a TV programme on the Ridgeway School of Adventure, which showed groups of managers being pushed to breaking point walking up a mountain in appalling conditions and being encouraged to swim beneath the keel of a boat in the icy waters of the Atlantic. That sort of approach is exclusive and, in my opinion, presents value to only a small percentage of participants.

It is possible to use of the outdoor environment very successfully for building teams and relationships and to some extent as part of a management training programme, but outcomes need to be clearly set and understood by all concerned, and the diversity present within the group needs to be managed effectively. Physical challenges have to be inclusive and creative, and planning and organisational challenges need to complement the physical components so that everyone has an opportunity to shine.

Here are some examples of team- and relationship-building events I've run that use the outdoor environment, but remain both challenging and inclusive:

1. *Footpath Survey Work*
    Objective: Learn about working in, managing and leading teams.
    A 3-day residential course in which a group from Loughborough University was asked to complete two projects:
    a. To work in teams and survey a section of footpath and assess the condition of footbridges, drainage, surfacing, erosion and outdoor furniture such as seats. The report would be presented to the National Park service to help it identify priority work. The groups had to research

and decide on the recording method and what standards to use, delegate roles, appoint leaders and so on, then compile and produce their report in conjunction with the other groups, who had been surveying adjacent paths. This task was real, useful and outdoors based and presented lots of opportunities for experiencing different roles and analysing the assets required for effective teamwork.

b. Teams were reformed and each of the new teams had to first plan, then construct, a boat out of a range of provided materials that would be capable of carrying an egg, unbroken, across one of the Lake District tarns. The planning and building phases were separate, and the groups were expected to have developed their teamwork skills following reviews of the previous exercise. The grand finale was the boat race – always fun and often quite exciting!

c. Each of the activities was followed by reviews that highlighted the learning that had taken place, and the expectation was to see observable changes in behaviour in relation to the participants' ability to work effectively in teams and in terms of developing their personal roles. Back at the university, academic work on managing tasks and teams complemented and put into context the practical experience gained.

2. *Filmmaking*

Team building event for a private company.

Brief: Make a 2-minute TV advert promoting adventure activities in the Lake District.

The project involved each team learning how to use a range of video equipment and put together a storyboard for a TV advert. The teams also had to identify activities that they could film, find locations and acquire the necessary skills to be able to undertake the activities. There were a number of clearly defined roles such as director, producer and so on, and these roles were interchangeable to provide a range of opportunities. Periodic reviews assessed progress from a technical perspective and also from a learning point of view, with tutors helping to pick out the key issues. At the end of the project, the raw film and final storyboards were sent to a local editing company, which worked overnight to edit and produce the films. This was followed by a grand 'premiere' to judge the winning advert. The results were very professional and provided a highly entertaining evening.

3. *Multiactivity Relationship-Building Event*

A series of multiactivity relationship-building events for the Olympic Delivery Authority (ODA) held in the years building up to 2012:

There was no formal training involved in these events – they were purely aimed at improving relationships between ODA staff and their delivery partners through an event called 'Lakes Challenge'. Participants were offered a range of activities varying from tough challenges that involved combinations of mountain biking, hill walking and canoeing, to softer outdoor activities such as archery, photography and pony trekking. Over the course of 2.5 days, the group mixed successfully on the activities and during social time, with the whole event drawn together with a final night dinner and prize giving. The event worked particularly well due to the hierarchical mix within each activity group, the fact that groups changed composition frequently and the excellent hotel facilities that allowed easy and effective socialising. Provision of activities to suit all ability levels was also a significant help. The participants raised an impressive amount of funding for charities local to the 2012 project. The final reason for the success of these events was the wholehearted support of the 2012 hierarchy, with the CEO regularly attending the events and setting a lead example.

4. *Health and Safety Team Build*

I've recently developed a new team-building event that has health and safety at its core. Team-building events can have many functions, but I believe that this is the first one set up to provide a combination of increased awareness of health and safety issues alongside a strong team-building function.

The course is run on a residential basis and is built around the concepts of free-thinking hazard identification (see Chapter 11), marginal gain, well-being and the other key components found in this book. The course contains some short, formal lectures, but the majority of the time is spent carrying out unfamiliar outdoor tasks that present genuine hazards. Identifying these hazards and managing them presents a significant challenge, and from a training point of view, there are many opportunities to examine and learn from the processes involved. Groups also directly experience some of the methods described in the book, including being provided with a specific diet during their stay along with longer-term advice, a chance to experience the positive benefits of exercise and time spent in the outdoors and the services of an on-site physiotherapist to replicate the workplace value of early intervention. Working in groups, there is an additional strong element of team building and the overall benefit of a powerful, shared experience.

Team-building events can and should have multiple benefits, but the following list illustrates the potential positive benefits that relate to our aim of reducing accidents, injury and ill health:

- Team-building events should, by definition, create stronger teams and better, more understanding relationships.
- They provide opportunities for improved well-being and can create a happier workforce.
- If targeted directly at a combination of health and safety awareness and team/relationship building, the benefits can be more direct.
- Closer relationships can bring about a more caring attitude between employees that can in turn foster higher levels of safety as people 'look out' for each other more.
- Team-building events can kick-start a person's desire to be fitter and healthier, and trigger positive changes in behaviour.
- Targeted events can build problem-solving skills and help people to manage changing circumstances from a health and safety perspective.
- Events can be used to identify workplace risks and help plan future health and safety policy.

- Building closer relationships in conjunction with the right content can help people to share personal problems that may be impacting their working lives.
- Reduction in stress levels.
- Improvement in communication skills, monitoring and reviewing techniques.
- Opportunities for a cross-hierarchical approach that can help establish good working relationships and embed the philosophy that health and safety are a shared responsibility.

Management actions:

- Ensure team-building events have very clear aims and are clearly integrated with longer-term organisational strategies. Share this information clearly with participants.
- Wherever possible, make every effort to involve personnel from across the workforce and the hierarchical range. This demonstrates and establishes management support, and from a health and safety perspective, it helps to embed the principle of shared responsibility.
- Look for team-building events that help to improve well-being and reduce levels of stress alongside other key objectives.
- Consider team-building events that focus on health and safety and use them as part of your policy of reducing accidents, injury and ill health.
- Ensure that health and safety are an integral part of every event and find some way of incorporating them as a topic. For example, this could be as simple as asking the group to draw up their own health and safety policy for the event.
- Be sure to follow up on each event internally, to maximise and consolidate learning benefits.

Team building can play an important role in helping improve health and safety, but very clear aims have to be established alongside an event follow-up that consolidates learning. Building strong teams makes an organisation more resilient and effective; makes a happier work environment and can reduce stress, bolster confidence and improve communication.

Creating good working relationships and effective teams are strong enablers that will help with the implementation of the strategies outlined in the book.

Printed in the United States
by Baker & Taylor Publisher Services